ROCKHOUNDING
Montana

Help Us Keep This Guide Up to Date

Every effort has been made by the author and editors to make this guide as accurate and useful as possible. However, many things can change after a guide is published—trails are rerouted, regulations change, techniques evolve, facilities come under new management, etc.

We would love to hear from you concerning your experiences with this guide and how you feel it could be improved and kept up to date. While we may not be able to respond to all comments and suggestions, we'll take them to heart, and we'll also make certain to share them with the author. Please send your comments and suggestions to the following address:

Globe Pequot
Reader Response/Editorial Department
246 Goose Lane, Suite 200
Guilford, CT 06437

Or you may e-mail us at:

editorial@falcon.com

Thanks for your input, and happy trails!

ROCKHOUNDING
Montana

A Guide to 100 of Montana's Best Rockhounding Sites

Third Edition

MONTANA HODGES

FALCONGUIDES

GUILFORD, CONNECTICUT
HELENA, MONTANA

For the author of the first edition of *Rockhounding Montana*, Robert Feldman, and his wonderful wife, Claudia. Had it not been for their years of hard work and dedication, many sites in Montana would remain silent.

FALCONGUIDES®

An imprint of Rowman & Littlefield
Falcon, FalconGuides, and Outfit Your Mind are registered trademarks of Rowman & Littlefield.

Distributed by NATIONAL BOOK NETWORK

Copyright © 2016 by Rowman & Littlefield
Maps: Alena Joy Pearce © Rowman & Littlefield

British Library Cataloguing-in-Publication Information available

Library of Congress Cataloging-in-Publication Data available

ISBN 978-0-7627-8162-1 (paperback)
ISBN 978-0-7627-9907-7 (e-book)

∞™ The paper used in this publication meets the minimum requirements of American National Standard for Information Sciences—Permanence of Paper for Printed Library Materials, ANSI/NISO Z39.48-1992.

CONTENTS

Map Overview

ACKNOWLEDGMENTS

First and foremost I would like to thank Mom and Dad. Had they named me Virginia, things would have been so different. A special thanks to Dad, who has inspired me to go outside and explore the natural world since I was a child. He brought science to my life and a love of open space to my soul. And a special thanks to Mom, who has given me the strength and ambition to follow my dreams and the attitude to make it all come true.

Thanks to the author of the first edition of *Rockhounding Montana*, Robert Feldman, who lent great help to the second edition, who became the inspiration for my PhD, and who was responsible for the geologic integrity of Falcon rockhounding guides. He really helped me keep my strata straight and he will be missed.

Also a due thanks to my geologic field assistants—Jessica McCartney, Amy Singer, Emilia Palenius, Taylor Grage, Tanis Harty, Joey Latsha, Matt Blancas, Kip Sikora, and Ted Stevens. It is nearly impossible to properly express my gratitude to the geology department at Cal State Sacramento, particularly to the department chair, Dr. Dave Evans, who assured all the support a student in the field could ever dream of, not to mention the proper equipment. He was also responsible for the help of technician Steve Rounds, the geologist who spent much time identifying my many mysterious rocks and minerals. Thank you, Steve—you rock.

Of course, the biggest thanks to all the rock hounds, without whom life would be so dull. Hopefully this book will bring many adventures and trips filled with happiness and, of course, rock hounds will want to pass it on to a friend. See you in the field!

—*Montana Hodges*

Map Legend

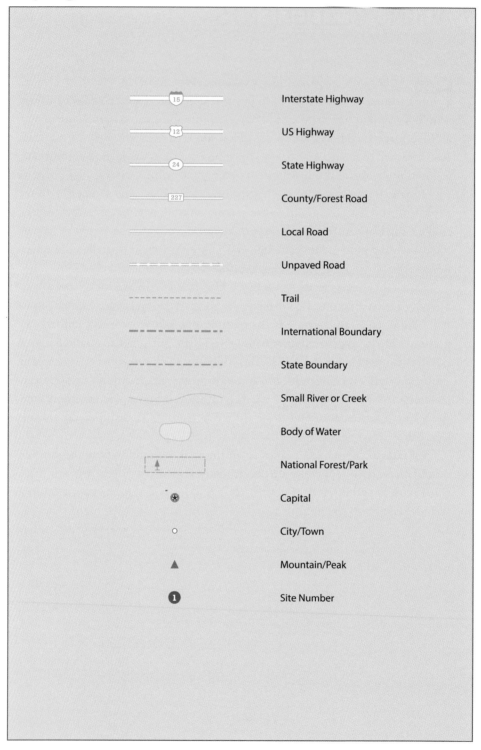

—15—	Interstate Highway
—12—	US Highway
—24—	State Highway
—227—	County/Forest Road
——	Local Road
-- -- --	Unpaved Road
- - - -	Trail
— - — - —	International Boundary
— - — - —	State Boundary
~~	Small River or Creek
⬭	Body of Water
▲	National Forest/Park
⊛	Capital
○	City/Town
▲	Mountain/Peak
❶	Site Number

INTRODUCTION

Only one state could be known as the treasure state, and Montana holds that honor. Possibly no reference could be more suitable for a land of such rich history, vibrant culture, and physical beauty, not to mention so many great rocks! Montana is home to gold caches, ruby-spiked rivers, miners' tales of precious cobalt sapphires clogging sluice boxes, and dinosaur beds so vast that paleontologists can't dig them up as quickly as they erode from the prairies. Montana is a state with legends of the "richest hill on earth," copper kings, Custer's Last Stand, flashy minerals in mysterious ghost towns, fossils hunts in eerie "badlands," trails traversing through petrified forests, and the chance to uncover ancient seas. Montana embodies the Wild West, a rugged outback, an isolated prairie, a bustling culture, and the chance for an adventure in what may be a rock hound's paradise.

Montana's geography of today is best described by such words as land, space, mountains, and sky. Only three states—Alaska, Texas, and California—have an area bigger than Montana's 147,046 square miles. The name "Montana" is derived from the Spanish word for mountain, and one glance at a state map reveals dozens of ranges in the west. Some mountain-crest elevations rise above 12,000 feet. Most of the state's population and tourism can be found in the valleys that divide these western ranges. Nestled in these valleys are historical towns such as Helena, Missoula, Butte, and Bozeman.

The mountainous western third of Montana contains a treasure trove of precious gems and minerals. Suitably, the state's motto of *Oro y Plata* is Latin for "gold and silver." Deposits of many metals, including gold, silver, and copper, are located throughout the western mountains and set Montana's roots as a major mining state. Garnets, topaz, and sapphire (the state's precious gem) erode out of their many deposits in the west.

On the other side of the Continental Divide, the line at which water on the east flows to the Atlantic Ocean and water on the west flows to the Pacific, the landscape fades into hundreds of miles of rolling hills and prairie grasses. Here, the Great Plains and the Yellowstone River lay uninterrupted under infinite sky. Only eroded desert areas known as "badlands" stand as testament to the extremities of plains weather and eastern living. Arctic air from the north influences the weather of the plains. Low-lying mountains and prairie

Tailings piles and old buildings, like these in Granite Ghost Town, stand as testimony to the mining craze in Montana's early history.

offer little shelter from the freezing wind of the winter and burning heat of the summer. But to an appreciative observer and the small percentage of the state's population who reside there, the lunar landscape is worth the setback. The mountains may be a source for the state's name, but the plains of eastern Montana are the inspiration for the state's most famous nickname: "Big Sky Country."

Meriwether Lewis and William Clark were the first recorded American explorers to officially travel across Montana during their US government expedition of 1805–06. The men represented the beginning of contact between the native tribes of Montana and the ambitious hunters to come. A goal of the expedition was to evaluate profit that could be made from land exploitation throughout the territory. At the time fur trapping was thought to be the most promising western enterprise. After Lewis and Clark spoke of the Montana territory as "richer in beaver than any other country on earth," trappers and traders flooded the region. The fur craze of the 1800s was the beginning of the end to the original lifestyle of Montana's Native Americans. Of course, it also didn't take long for Montana's newest inhabitants to discover metal deposits and begin the quest to remove the earth's minerals.

Eastern Montana is responsible for the state's nickname, Big Sky Country.

The fur craze, mining craze, and the new settlers pushed the Native Americans farther from the necessary lands for survival. The Sioux and Cheyenne fought the last successful battle by natives in 1876 at the Battle of Little Bighorn. As the desirable land became more scarce and after a series of bloody conflicts, the last native people were pushed onto reservations throughout the state. Today several reservations occupy Montana territory, and Native Americans continue to have a strong physical and cultural presence in the state.

Montana entered the union in 1889, becoming a young forty-first state of the United States. The Montana of today is based less on mining natural resources and more on agriculture. Agriculture is one of the state's primary industries, but it is more struggling than robust. The population of the state has always managed to remain small, recently topping 1 million people. The state may be one of the largest in area, but it ranks forty-fourth in population.

Home to some of the nation's most popular national parks, from the ice-capped peaks of Glacier to the spurting hot springs of Yellowstone, Montana encompasses vast areas of protected lands. Crystal lakes, undisturbed forests, clean air, and the largest variety of mammals of any state are just a few of the traits that fulfill the desire for unspoiled landscape. A state with three times as many cows as people, it represents a different world inside a busy country. Montana truly is America's outback.

MONTANA'S GEOLOGIC INTRODUCTION

The experience of rockhounding in Montana is greatly enhanced by an understanding of the state's geology. Most rock hounds aren't geologists, and most geologists don't write books on rockhounding. Yet they are often tied together in interests and the appreciation of understanding the natural world and all its complexities.

A plethora of reasons surround each geologic feature of the state, many of which are best explained properly by a geologist. There are many books that describe Montana's geology, one most highly recommended for the journey is *Roadside Geology of Montana*, by David Alt and Donald W. Hyndman. The book is an excellent companion to have on your rockhounding journey.

It is not the bottom of an ocean that people expect to find when walking across the high plateaus of Montana. Such items as seashells, coral, and other marine life aren't what initially come to mind when we think of the land-locked northwestern states. Yet, there it is, a giant clam in the high desert, a shark's tooth on the side of a mountain! Marine fossils from the Idaho border to the Dakota line. Wait, and underneath the clams are terrestrial dinosaurs and plants, precious gems and mammoth bones. Even Lewis and Clark wondered why petrified skeletons of large "fish" they observed would be so far inland (one was a plesiosaur).

In a nutshell, Montana's geologic story is a tall tale. Scholars continue to understand more about the complex history of the state, but still there is much to learn. The oldest rocks in Montana, the basement rock, have been dated as far back as 3.2 billion years, and since that time, life-forms have come and gone, seas have filled and drained, mountains have been forced up, volcanoes have formed and erupted, glaciers have passed by, and the entire North American continent has changed shape.

There are some important geologic events in Montana to be familiar with. The basement rock may date from more than 3 billion years ago, but much of the rock you will come across is not this old. During Precambrian time (see the Geologic Timetable below), Montana was covered by an ancient sea. Sediments of this shallow sea produced the Belt series rock on top of the basement rock. For several hundred million years, these fine-grained deposits of sand and silt accumulated and eventually hardened into the slightly

Western Montana is known for its mountains, valleys, and world-class trout fishing.

altered Belt series sandstones, and shales, that are visible throughout western Montana. Mudcracks, ripple marks, and other evidence of the shallow sea and even some primitive life such as algae can be viewed on these rocks today. These rocks deserve mention because you'll get to know them as you rockhound across Montana.

Transitioning to Cambrian time, shallow sea, freshwater, and terrestrial deposits continued to accumulate on top of the Belt. The first evidence of complex life appears and explodes. During the periods of the Paleozoic era that followed, thick beds of organic deposits in the east would eventually become some of the state's major gas and oil caches. Shallow seas continued to transgress and regress across Montana into Mesozoic time, with a particularly notable inhabitant—the dinosaurs. The end of the Mesozoic era and the beginning of the Cenozoic era is marked by the abrupt disappearance of the dinosaurs. This dramatic change 65 million years ago is often referred to as the "K-Pg" boundary (an abbreviation of the boundary in time between the Cretaceous and the Paleogene). It is the subject of much controversy and study. There are only a few places in the world where the layer of sediments from this period of time can be viewed with the naked eye. Garfield County, Montana, is home to several of them.

As the Cenozoic moved on, massive vegetation deposits were buried in Montana, leading to the coal and lignite caches found in central and eastern Montana. In the west the Rocky Mountains began to form as the North American plate collided with the floor of the Pacific Ocean. Massive fault zones were created, batholiths emerged, and mountain-building events were in full throttle. Meanwhile, the east remains relatively quiet. A slight eastern uplift occurs as the Rocky Mountains formed, creating exposures of many of the fossiliferous layers we rockhound today. The Yellowstone hot spot erupted for the first time about 2 million years ago. The climate of Montana shifted from tropics to ice ages, mountains continued to form and erode, animals evolved, glaciers carved valleys, and eventually the landscape we know today came to be. Now it is time to go rockhounding across Montana—a journey of 3 billion years.

These processes are in many ways what have created the minerals, fossils, and gems that we search for today. When at all possible in this book, a geologic description has been included with each site to explain how specimens came to be and their relative age. Yet the most important aspect for the honorary geologist-in-training (which rock hounds are) is to ask questions about the rocks you find. "Where did this come from? How old is it? Why was it there?" "Oh dear, poor snail, how did he die?" "How does garnet form?" Once you've adapted to asking and answering questions, you can pull out a sapphire you mined personally and dazzle friends with your mineralogical expertise (field guide to gems and minerals required) or explain how your spectacular ammonite may have coexisted with the last dinosaur! These joys of discovery and provocative ways of learning are what will truly bring the pleasure of rockhounding to the soul. So prepare to quest for scientific knowledge. Soon the kids will be asking, "How old is a billion? Older than you?"

MONTANA'S STATE GEMS

Many states have an official gemstone—Montana has two. In 1969 agate and sapphire became the official state gemstones. These state stones are unique due to their abundance where just about anyone has the opportunity to collect them.

Montana Sapphire

Montana's sapphires weren't always a highly prized gem. During Montana's gold-rush era, sapphires were cursed for clogging sluice boxes and getting in the way of gold mining. Later, as their value was learned, some gold operations, such as Montana's famous Yogo Gulch, were adapted to mine specifically for the sapphires.

SIMPLIFIED GEOLOGIC TIMETABLE

Era	Period	Epoch	Major Event
Cenozoic	Quaternary	Recent (10,000)	Ice Ages
	Paleogene	Pleistocene (2,000,000) Miocene (23,000,000) Oligocene (33,000,000) Eocene (55,000,000) Paleocene (65,000,000)	Age of Mammals
Mesozoic	Cretaceous (150,000,000) Jurassic (200,000,000) Triassic (250,000,000)		Age of Dinosaurs
Paleozoic	Permian (300,000,000) Pennsylvanian (320,000,000) Mississippian (350,000,000) Devonian (420,000,000) Silurian (440,000,000) Ordovician (500,000,000) Cambrian (550,000,000)		Age of Invertebrates
Precambrian	Proterozoic (2,500,000,000) Archean (4,000,000,000)		Earliest Life (3,200,000,000)

(Numbers refer to years before present)

Miners work at a Yogo sapphire mine.

Raw naturally cobalt-blue Montana sapphires

Sapphires are a form of pure gem–quality corundum occurring in many colors, notably blue. In Yogo Gulch sapphires naturally occur in a rare deep cornflower blue. Montana's Yogo sapphire was the only North American gem to be included in England's "Crown of Jewels." The Yogo Mine is located on private land, off-limits to collecting, but there is barely a stop in Montana where a Yogo isn't featured on display or for sale. These unique gems are some of the most beautiful in the world; many sapphires found elsewhere are heat treated in an attempt to mimic the natural cobalt beauty of a Yogo.

For a detailed look at the history and geology of Montana sapphires, consult *Yogo: The Great American Sapphire* by Stephen M. Voynick.

Outside of Yogo, there is still opportunity for the public to mine sapphires. This satisfying treasure hunt is highly recommended. There is nothing quite like wearing a sapphire you found yourself.

Montana Agate

There is so much to be told about agate. A chapter alone could be written on the various patterns and colors to be found in the magical stones, let alone the geology, history, and varieties. So it's a good thing that Montana has Tom Harmon (who owns the Agate Stop—Home of the Montana Agate Museum, with his wife; see Appendix D), one of the original agate masters and author of the must-have book *The River Runs North: A Story of Montana Moss Agate*. This book covers everything to know about agate and is full of some of the most stunning pictures of agate to be found.

Agate is a variety of quartz, more specifically a form of chalcedony that is very closely related to opal. In eastern Montana, agate is relatively common and was used by natives for thousands of years to fashion tools. Of course, they had impeccable taste, because the agate occurring in Montana is one of the more attractive and certainly a unique chalcedony variety to be found. The stone is popular due to its translucency, polishing capabilities, and wide variety of color and patterns. It can be found easily and has potential to be fashioned into jewelry, carvings, and other ornamental pieces.

"Montana moss agate" is the most sought-after variety of agate in the state. "Moss" refers to dendritic patterns that can occur in the translucent stones. Various images and beautiful landscapes can sometimes be seen in the patterning of the moss agate after it has been cut and polished. At any time of low water, agate can be commonly found throughout the gravels of the Yellowstone River. An eye must be developed to find a piece, as they often look like plain river rock due to their deceiving dull crusts. Hints of translucency can be seen in the sunlight—an instrumental tool to the search. A trip to Montana isn't complete without finding a beautiful moss agate. From

Author Montana Hodges and agate master Tom Harmon collect agate on the Yellowstone River in late summer.

pebble-size nodules to crystal-filled geodes and limb casts of trees, there is plenty to be found—perhaps the stunning piece for your next bola tie.

Montana agate is often collected in the gravels of the Yellowstone, but it was formed long before the time of the river. Most Montana moss agate is believed to have been created about 60 million years ago during volcanic activity in what is today the Yellowstone Park area. Agate can form as volcanic flows cool; in some way or another, water rich in silica would flow through lava and fill cavities. Perhaps the liquid silica would fill just an air pocket, or maybe a cast of a buried tree limb, and eventually harden into agate.

ROCKHOUNDING

Rockhounding is a hobby for a special person. To some, the "hunting" of rocks can seem quite bizarre. After all, if it were just rocks someone was after, they're everywhere. Rock hounds collect many different kinds of rocks for many different reasons; some collect fossils, or rocks, or minerals, and some don't keep anything at all. To many, rockhounding can be much more than finding rocks. It can take you to locations beyond the standard roadside attractions and allow you to explore areas of America that many tourists never see. Rockhounding pushes you through towns where outsiders seldom come, on roads without pavement, past cattle trails, to ghost towns, to abandoned quarries, to forgotten mining districts, and to mountaintops. Perhaps rock hounds are the last of the American explorers.

The best part is that anyone can be a rock hound. Whatever your flavors, from ammonites to micromounts, whether you're looking to collect for an hour or a month, stay close to the car, or prospect the Rockies, there is a site in Montana for you. And maybe you'll learn something new or bring home a gigantic sapphire. Rockhounding is a great family sport. You don't need television, junk food, soda, or a scoreboard (although if you bring all of these along, the group may last longer). I have never run across an educational pastime as entertaining as looking for rocks. So bring the kids. The more kids the better; they're closer to the ground—they find better stuff! It's treasure hunting with an opportunity to learn. Here is your chance to make science fun and tell the kids that they *can* touch things.

Collecting Regulations and Etiquette

Public lands make up much of Montana's landscape, creating millions of acres open to people for exploration. Rockhounding is allowed on several forms of public land. In general, rockhounding is allowed on Bureau of Land Management (BLM) and forest service land, with restrictions. Rules and regulations vary by location and could change at any time, so always inquire before collecting.

It may seem like a pain, but visiting every BLM office and ranger station is all part of the experience. Oftentimes there are multiple benefits to stopping at the office. Aside from checking land status, the staff can be helpful. They may give you great advice on where to rockhound, camp, hike, or lunch. An insider's tip could become the highlight of the trip.

There are some basic regulations to be familiar with while collecting on public land. Collecting or disturbing vertebrate fossils on government property is not allowed without a proper permit. This rule is strictly enforced. Also, all findings are subject to the Federal Antiquities Act of 1906, and anything of historical value should not be removed. Anything of archaeological nature is protected under the Native American Graves Protection Act and should not be removed. This means that an arrowhead made of agate is not legal to collect. It may have been agate first, but it is an artifact now. The good news is that any rock, gem, mineral, or fossil that you do find while abiding by the rules is yours to keep (like that 2-ounce nugget!).

USDA Forest Service Land

Rockhounding is permitted on national forest land, although rules and regulations vary by location. These guidelines are ever-changing, so always contact the local forest service office for an updated map (see Appendix C) of the area in which you are collecting and inquire about the land status.

In general, rockhounding and gold panning can be done without a permit, fee, or special permission. Any kind of mechanical equipment and blasting are prohibited. All activity must be done without significantly disturbing the environment. A good example of what is against regulations would be digging excessively into tree roots for crystals. Many trees die from this activity, creating safety hazards and an ugly public forest.

As on all federal land, collecting vertebrate fossils is not allowed and became officially illegal in 2009 with the Paleontological Resources Preservation Act. Collecting of vertebrates, meaning any animal with a backbone (i.e., fish, mammals, dinosaurs) is punishable by law. It is still legal to collect invertebrates, such as fossil shells and plants. Material that is allowed to be collected must be taken for personal use with no commercial purpose. Size regulations vary by location. Exact collecting limits are not generally specified, but quantities are required to stay small. A few rocks are fine, but enough to make a new pathway through your rose garden probably requires a permit. A good example of a variation to the standard rule is the Gallatin Petrified Forest. A permit is needed to collect petrified wood in the boundaries of Gallatin Petrified Forest, and specimens collected are limited to 20 cubic inches per person per year.

Bureau of Land Management Land

The same rules apply to BLM land, although in this case, there are specified limitations. No more than 25 pounds per day, at a maximum of 250 pounds per year, can be taken per person from BLM land. This is still a generous offer, but remember that all collecting must be done for personal hobby use, just as on national forest land.

Other Federal Land

Collecting within national parks, monuments, wildlife refuges, and military land is not allowed without a permit for a qualified party. These rules are highly enforced, especially in areas with great geologic significance.

State Land

State land may be open or closed to the public. Always inquire for status before entering state land. When state land is leased, it may function similarly to private property, and trespassers can be prosecuted if they do not obtain permission from the lessee. If it is open to the public, rockhounding is still restricted and collecting of anything on the land probably requires a permit.

Tribal Land

Tribal land functions similarly to private land. In order to enter or collect, you must obtain permission through the Office of the Superintendent of the Bureau of Indian Affairs associated with the reservation in question, and if approved, the landowner of the particular location.

Private Land

Private land is simply private. Unless permission is obtained from the landowner, private land is not accessible, much less available, for collecting. Never assume that because there aren't signs or fences that the land is open for collecting. In some areas of Montana, you can drive for miles through private land and never see a fence (you may see the landowner's cows in the road, however). If permission is obtained from the landowner to collect on the property, anything you find belongs to the landowner unless the landowner relinquishes ownership. It is legal to collect vertebrate fossils on private land if the landowner permits, and many commercial dinosaur digs in Montana offer the opportunity.

Mining Claims

Mining claims function similarly to private land, even when they are located on public land. Often the public is allowed to pass through the physical property but, unless permission is obtained from the claim owner, disturbing or removing material from the claim is not allowed. Respect others' claims and be on the lookout for their postings. Mining claims are generally posted on small poles or tree trunks. To avoid trespassing, it is always best to stay off and away from anything you suspect is claimed.

HOW TO USE THIS BOOK
Maps

The regional maps found in this book have latitude and longitude lines to help find the general location of the rockhounding sites and to assist with GPS usage. The importance of having a detailed map cannot be overemphasized. In many parts of rugged Montana, an atlas isn't sufficient. A forest service, BLM, or other kind of large-scale map of backcountry roads is needed for many of the sites. A wrong turn on an unlabeled forest road could lead to hours of despair. Maps can be ordered online in advance (see office locations in Appendix C) or purchased from various offices during their business hours.

Using the Global Positioning System

New to this book is the Global Positioning System (GPS) coordinates at each site. A GPS is a wonderful tool to have, and many new vehicles come equipped with one. The system works off satellites and gives the degree coordinates of latitude and longitude of its location with average accuracy of around 20 feet. A fraction of a degree can be read as minutes and seconds or as a decimal. The inexpensive handheld GPS is popular with hikers and is highly recommended for any trip into the wilderness.

When possible, the coordinates listed for the sites in this book are at the most productive location. If the site involves you parking and exploring, the reading may be from the parking area. The coordinates listed are intended to get you within a few hundred feet of the area. Once you arrive, you then must use your rockhounding skills and luck to find the treasure. Also remember that exposed rocks and productivity change consistently with weathering and erosion, sometimes drastically, so exact locations are variable. Specific GPS coordinates are not provided for sites with mineral deposits or material that are spread out over wide areas or are sporadic.

Seasons

The recommended time to visit each site is often described by season. General terms are used instead of specific months because of the unpredictable snow-melts, early winters, etc. Spring refers to the time after the snow melts enough to drive on the roads, which could occur anytime from April through June depending on the year and elevation. Summer is when snow is no longer falling. Fall is when the temperature reaches freezing but the roads are still accessible. It is important to inquire locally about road conditions and check local weather forecasts before trekking out into the wilderness, especially at high elevations.

WEATHER

Montana's weather can be unpredictable and extreme. Thunderstorms are common during the summer months, so be prepared for the safety hazards associated with this. Each site was visited during good road conditions, and consequently, the descriptions of road conditions apply to nice weather. Recent storms, rain, snow, debris, and many other things can affect accessibility. A road that may be listed as suitable for any vehicle could become a high-clearance four-wheel-drive jeep trail when muddy. Use your best judgment, or call it a day when the dark clouds roll in.

WHERE TO STAY

A Montana map can be deceiving. Many names on the map that appear to be towns turn out to be fields (or abandoned train stops) when you get there. This is a common problem with visitors to the more remote areas of the state. Bold print doesn't necessarily mean anyone will be there, much less that there is a hotel. It is important to research your stops and secure a place to stay well in advance, whether it's a campsite, RV space, or motel, all of which can be booked solid during summer months.

SAFETY

Safety in rockhounding involves, as it does in most outdoor hobbies, exercising common sense and being prepared. The weather in Montana can be unpredictable, and sunny weather can quickly switch to flash floods and lightning. Montana is also home to rattlesnakes, grizzly bears, huge mosquitoes, deerflies, gnats, and many other dangerous and pesky creatures. Always make your presence known in bear country and watch where you put your feet, especially in areas frequented by rattlesnakes. Take a can of bug repellent with you wherever you go.

Bring the proper equipment and protection to each site: safety goggles, gloves, and a first-aid kit. Be prepared for getting lost or stranded. There is nothing more embarrassing than being stuck in a pile of snow on isolated Sweetwater Road outside of Dillon, Montana, at 10 o'clock at night without a flashlight or shovel. Basic supplies are necessary, and cell phone reception cannot be relied upon, especially outside of big cities.

One of the biggest safety hazards that rock hounds face is abandoned mines. The motto is "Stay out—Stay Alive" and it couldn't be more aptly summed up. Mines are extremely dangerous and should be avoided. Many minerals occur in the tailings piles around mines, and it would be a shame for the poor judgment of those who enter these dangerous mines to risk making collecting in the nearby areas restricted. Never go inside an old mine. Everything good has already been removed. If a site has known shafts, that fact will be mentioned, and children should skip these areas.

Most public land in Montana that is open to rock hounds is also open for hunting during certain times of the year. Always inquire about hunting season, which is generally in winter when the snow is too heavy to rockhound in remote areas anyway. If you do visit a site during hunting season, it would be wise to wear an orange safety vest and other bright colors while in wooded areas.

TEN VISTAS FOR YOUR MONTANA JOURNEY
Glacier National Park

You can't collect here, but people from across the world come to view the mountains and glaciated peaks of northwestern Montana's Glacier National Park. The park preserves more than a million acres, including some of the most breathtaking landscape. Lush, green valleys carved by the ice ages sit below towering mountains. In summer hundreds of species of wildflowers bloom in these picturesque lowlands. Dozens of lakes line the mountains, and approximately fifty glaciers dot the mountains. Some of the glaciers are visible from the road, and others can be seen through more than 700 miles of hiking trails. You can also find some of the oldest fossils in Montana, Precambrian stromatolites (algae) fossils, within the trails and roadcuts. The park is home to almost every kind of large mammal native to the United States and more than 200 bird species. While there don't forget to travel on the "Going-to-the-Sun Road," a scenic route through the park that crosses some of America's best photographic opportunities. The park is generally open from late May to mid-October.

Getting there: The park is located in northwestern Montana on the US–Canada border. The park and its headquarters can be accessed from US 2 in the west; there is also an entrance from US 2 on the eastern side of the park. For more information visit www.nps.gov/glac.

Yellowstone National Park

Although most of Yellowstone National Park is actually located in Wyoming, three of its five entrances are in Montana, and the park represents much of the state's tourism. Yellowstone is visited by people from around the world, and it's home to one of this country's largest wildlife sanctuaries. The park is also one of the world's largest active volcanoes. A rock hound can really appreciate this unique chance to view the magnificence of the volcanic process in one of the grandest thermal basins. Spurting geysers, terraced hot springs, and even boiling mud can be viewed easily. Wildly colored thermal pools and "screaming" vents of heated gases endure gloriously for generations of photographers. Mammoth Hot Springs, Old Faithful, and the Grand Canyon of the Yellowstone are some of the park's more famous destinations. Yellowstone is also home to many species of wildlife, including the once-endangered gray wolf, and the park is noted for being one of the most successful wildlife preserves in the world. More than 2,000 bison roam the area, and they often stroll leisurely along the road. Beyond the pavement more than 1,200 miles of trails cross the park's many wonders. The park is open for automobile travel from May through September. Limited access can be gained by other modes of transportation or limited automobile access during the off-season.

Getting there: The park is located in the southwest corner of Montana and has five entrances. The headquarters are located in Montana at the north entrance through Gardiner on US 89. The northeast entrance to the park is through Cooke City, and the west entrance to the park is located in West Yellowstone off US 191. The park can also be reached from the towns of Jackson and Cody in Wyoming. Visit www.nps.gov/yell for more information.

Beartooth Pass and the Beartooth Mountains

The Beartooth Highway (US 212) east of Cooke City is a scenic 64-mile stretch of road that connects Red Lodge with Cooke City and the northeast gate to Yellowstone National Park. The Beartooth Pass crests at more than 10,000 feet, an elevation well above the tree line, and this is one of the most beautiful drives in the state. Here, high mountain country and the occasional summer snow create wonderful photographic opportunities. The Native Americans of the area called the climb the "trail above eagles."

The Beartooth Mountains surrounding Cooke City are noted for their unique geological features. Active glaciers of any significant size no longer exist here, but a few small remnants are present. Of these, Grasshopper Glacier is perhaps the most interesting. Apparently, at the time glaciers were active in these mountains, swarms of locusts became trapped in the heavy snows and were frozen. Eventually they were buried by additional snowfalls and were incorporated into the glacial mass. Many layers of these creatures can be viewed in this glacier, particularly during the late summer months when the ice is better exposed. It is a hike of considerable distance to the glacier though, since the wilderness designation of the area does not permit motorized travel, but it makes a most enjoyable outing.

Getting there: US 212 connects Red Lodge with Cooke City and can be reached either heading south out of Red Lodge or from Cooke City just above the northeast entrance to Yellowstone National Park. Visit www .beartoothhighway.com for more information.

The Butte Area

The Butte area is probably the most exciting place in Montana for attractions that rock hounds can really appreciate. The World Museum of Mining and Hell Roarin' Gulch just outside the town limits of Butte are built on the grounds of the Orphan Girl mine. The museum has many displays on mining history and a reconstructed 1800s mining town with more than fifty buildings known as Hell Roarin' Gulch. Yet, the "gold mine" of Butte's special attractions is the Mineral Museum on the grounds of Montana Tech University. Along with geologic history and information, the museum has 1,500-plus minerals on display, including a fluorescent-minerals room, 400-pound quartz crystals, and what may be the largest gold nugget ever found in Montana. Also, while in the area don't forget to stop by the free Berkeley Pit Viewing Stand to view the consequences of excessive rockhounding. The giant crater stands as testimony to what was one of the world's largest open-pit copper mines. In the Early 20th Century the Berkeley Pit, also known as "The Richest Hill on Earth," produced more than $2 billion worth of gold, silver, copper, and zinc.

Getting there: The World Museum of Mining and Hell Roarin' Gulch are located at 155 Museum Way in Butte. The Mineral Museum is located on the grounds of Montana Tech inside the Museum Building at 1300 West Park St. The Berkeley Pit Viewing Stand is located on the eastern end of Park Street. All attractions are open year-round; inquire about exact times (mining museum.org).

Bighorn Canyon

Beautiful Bighorn Canyon and the Pryor Mountains west of it are located in the high desert on the Montana-Wyoming border. The Bighorn Canyon National Recreation Area is a wonderful stop for water sports, walleye fishing, and handsome scenery. It was named for its deep canyon walls along 71-mile-long Bighorn Lake. There is much wildlife in the area, from world-class fishing in the Bighorn River to wild-horse viewing in the Pryor Mountain horse range just outside it. Keep an eye out for bald eagles, one of hundreds of bird species to be seen in the canyon. Rockhounding is not allowed within the canyon, so include a stop at the Pryor Mountains (see Site #60) to collect some fossils.

Getting there: Bighorn Canyon, although in Montana, is best accessed through WY 37 north of Lovell, Wyoming. The canyon can also be reached from Montana by driving about 40 miles south of Hardin on MT 313. Visit www.nps.gov/bica for more information.

Fort Peck Dam Interpretive Center and Museum

Fort Peck Lake is home to one of the largest earth-filled dams in the world. New to this area is the impressive Fort Peck Dam Interpretive Center and Museum (www.fws.gov/refuge/Charles_M_Russell/visit/visitor_activities/ FPIC.html), located just below the dam. The center works together with Fort Peck Paleontology Inc. to host one of the most thrilling dinosaur displays in the state. Upon entrance to the enormous building, a life-size fleshed-out model of the Tyrannosaurus rex known as Peck's Rex greets visitors. Behind the T-rex, gigantic windows overlook the sublime lake. The museum holds many other paleontology displays along with exhibits on Montana history and wildlife and Fort Peck.

Getting there: The museum is located in Fort Peck just south of the dam, between the powerhouses and campgrounds.

Museum of the Rockies

The Museum of the Rockies is located on the grounds of Montana State University (MSU) in Bozeman. The museum has displays on many aspects of northern Rockies history, with a special emphasis on geology and detailed geologic exhibits with beautiful fossil specimens. The paleontology displays include the skulls of some of Montana's most famous species of dinosaur. Also in the museum are a planetarium and many exhibits on Montana history, including a large outdoor Lewis and Clark display. The Museum of the Rockies is open year-round; inquire about times and holidays (museum oftherockies.org).

Getting there: The museum is located in Bozeman on the MSU campus at South 7th Avenue and West Kagy Boulevard.

Gallatin Petrified Forest

It is a good hike of several miles round-trip to go there, but the Gallatin Petrified Forest north of Gardiner is worthy of the climb (see Site #55). Here an ancient forest that was once covered in ash, which turned much of the original tree material into petrified wood, has slowly been exposed through thousands of years of erosion. Magnificent tree trunks can be seen, and with a permit small amounts of material can be legally collected. Some of the material is of exceptional quality, but the highlight is by far the journey into the forest of the past via a beautiful mountain valley.

Getting there: The trailhead for the Gallatin Petrified Forest is located 16 miles north of Gardiner off US 89. Once you pass the small town of Miner, head 10 miles west on Tom Miner Basin Road.

Sun River Canyon

One of Montana's most secluded gems, the isolated Sun River Canyon is an easy escape from the ordinary. The canyon isn't off a road leading toward any popular destination, and most likely it is out of your way, but the atmosphere alone is worth the journey. The drive from Augusta to the canyon is stunning. Rolling high prairies and some of the biggest and clearest sky of the state lead to a dramatic shift of the surprising mountain range ahead. Within the range the Sun River and amazing limestone cliffs of the canyon await. Large Gibson Reservoir above the dam has beautiful sandy beaches and excellent fishing opportunities. Bring binoculars to observe the overwhelming amount of mountain wildlife. Hiking, camping, fishing, and rockhounding are all excellent at this location.

Getting there: The canyon is located 15 miles north of the small town of Augusta (west of Great Falls), on Sun River Road.

Makoshika State Park

Makoshika is Montana's biggest state park and by far the most interesting one. The name is from the Sioux language for "bad land." The park preserves more than 11,000 acres of eroded earth deemed badlands. Being anywhere in the park is like being in another world. Badlands are soft sandstones and shale formations sculpted by weather. The highly eroded high desert and plains formations look much like a landscape of a science-fiction novel. Giant sandstone boulders rest upon thin columns of earth, waiting to fall. Wash beds, alluvial fans, and crumbling hills of sands are interrupted only by the occasional coulee, spring, or desert vegetation. At sunset the sage-tinged placid landscape fades to colors of pink, orange, and red.

Many dinosaur finds, including several of significant scientific value, have been discovered in the Cretaceous Hell Creek Formation in the park. A small visitor center and museum located just within the gate to the park house displays on the geology and fossils of Makoshika. The park is open year-round, weather permitting.

Getting there: The entrance to the park is located 1 mile south of Glendive at the end of Snyder Avenue.

Sites 1 and 2

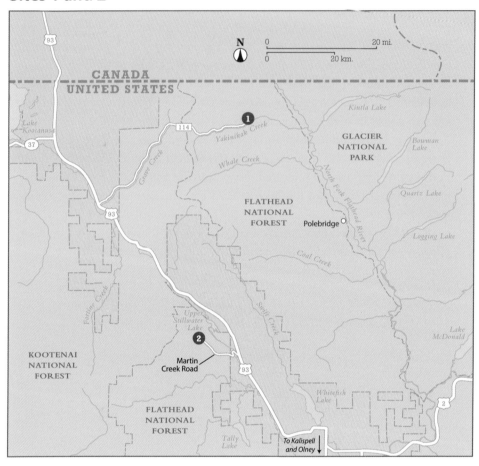

CANADA
UNITED STATES

N

0   20 mi.
0   20 km.

Lake
Kootanusa

Kintla Lake

GLACIER
NATIONAL
PARK

Bowman
Lake

Yakinikak Creek

Whale Creek

Grave Creek

FLATHEAD
NATIONAL
FOREST

Polebridge

North Fork Flathead River

Quartz Lake

Coal Creek

Logging Lake

Fortine Creek

Upper
Stillwater
Lake

Swift Creek

Lake
McDonald

Martin
Creek Road

KOOTENAI
NATIONAL
FOREST

Whitefish
Lake

FLATHEAD
NATIONAL
FOREST

Tally
Lake

To Kalispell
and Olney

1. Whitefish Fossils

A fossilized horn coral is found in a loose piece of limestone.

See map on page 22.
Land type: Mountains, roadcut
GPS: N48° 56.31900' / W114° 32.94660'
Best season: Summer
Land manager: USFS, Flathead National Forest
Material: Marine fossils
Tools: Rock hammer
Vehicle: Any
Accommodations: Camping and RV parking in Flathead National Forest; camping, RV parking, and motels in West Glacier
Special attractions: Glacier National Park is home to glaciated peaks and some of the oldest fossils in Montana.

Finding the site: This site is a tough one to find; a good forest map will be needed. The location is at a high elevation not far from the Canadian border; local inquiry should be made if plans are to visit in either spring or fall.

From Polebridge drive about 15 miles north on Outside North Fork Road (MT 486) and turn left onto Trail Creek Road. Drive 5 miles and turn right onto Thoma Creek Road (FR 114A). Drive 1.2 miles up Thoma Creek Road to the site. It is difficult to spot the limestone outcrops on the right side of the road because the mountainside is covered in lush vegetation. Subsequently, mileage and exploration is very important. Also keep an eye out for pieces of loose limestone near the edge of the road that might contain small fossils.

Rockhounding

The geology of far-northwestern Montana consists almost entirely of Belt rock and occasional diabase sills. In the area north of Polebridge, high in the Whitefish Mountains and within walking distance from the border with British Columbia, there are unique deposits of limestone. Some of the limestone contains horn coral and brachiopod fossils. The coral is quite large, more than an inch in diameter, and is fairly well preserved. Yet, diligent searching is required to locate both the site itself and fossiliferous rock. Splitting the hard limestone to expose or remove an intact fossil is very difficult as well. The best fossil hunting is likely done by investigating loose pieces of limestone beneath or near the roadcuts.

2. Kalispell Pyrite

See map on page 22.

Land type: Mountains

GPS: N48° 33.91140' / W114° 40.45200'

Best season: Late spring through fall

Land manager: USFS, Flathead National Forest

Material: Limonite, pyrite

Tools: Rock hammer

Vehicle: Any

Accommodations: Camping and RV parking in Flathead National Forest; camping, RV parking, and motels in Kalispell

Special attractions: Flathead Lake and Glacier National Park. The Northwest Montana Historical Society Museum in Kalispell has displays on local history.

Finding the site: From US 93 north of Kalispell, just before the town of Olney and about 0.5 mile after the Girl Scouts Camp, turn left onto Good Creek Road. Drive 3.3 miles and turn right onto Martin Creek Road. Drive 4.3 miles and park on the right at FR 910A. If the gate is closed, walk 0.5 mile up FR 910A to the sharp corner and several outcrops of blue-green rock in the mountainside. Look for evidence of previous digging.

Rockhounding

The Belt rock in this particular area produces some quality pseudomorphic limonite after pyrite cubes. Originally formed as pyrite, the limonite occurs as a secondary material due to oxidation of the original pyrite. Small samples of pyrite cubes can be found in the greenish–blue rocks just before the bend; but the limonite after pyrite cubes, which are up to 1½ inches in size, are found in the rock outcrops just at, and slightly past, the corner. The outcrop has been highly worked so you may have to expand your search from this starting point. The biggest cubes are found by breaking pieces loose from the parent rock and hoping for a lucky split. Small pieces can be found in the loose rocks at the bottom of the outcrop. The cubes tend to easily "pop" free from the belt rocks.

Sites 3–6

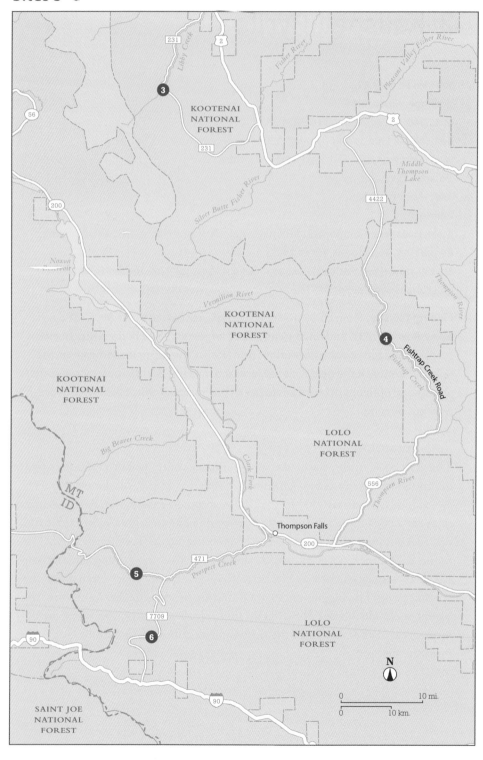

3. Libby Creek Gold

See map on page 26.
Land type: Mountains, creek
GPS: N48º 06.868' / W115º 32.991'
Best season: Late spring through summer
Land manager: USFS, Kootenai National Forest
Material: Gold, garnet
Tools: Gold pan, trowel, fleck flask
Vehicle: Any
Accommodations: Camping and RV parking at Howard Lake; camping, RV parking, and motels in Libby
Special attractions: The Heritage Museum in Libby has exhibits on mining, forestry, and local history.

Finding the site: From Libby drive 13 miles south on US 2. Turn right onto Libby Creek Road (FR 231) and drive 10 miles and left over a small bridge to the main parking area on the right. Large signs posted near the restrooms display a map of the area open for recreational panning and, more importantly, the area that is private and off-limits. The creek is located on one side of the parking area. Along with an informative website, a physical map is available at the Libby Chamber of Commerce and is highly recommended. The staff can also tell you about different gold-prospecting stores in the area if equipment needs to be purchased. For more information visit www.libbymt.com/areaattractions/libbycreekgold.htm.

Rockhounding

Mining began in the Libby area with placer deposits worked for gold in the early 1880s. The oldest lode mine was developed around 1887, and since that time mining has been sporadic and apparently not extremely profitable. Most of the mines were located about 20 miles south of Libby in the tributary valleys along the east slopes of the Cabinet Mountains and south of Troy (to the west of Libby). Unfortunately, these mines and tailings offer little opportunity to the rock hound as most are private. Despite the lack of mine tailings, the Forest Service has set aside this wonderful site for the public to pan out their own gold. Unlike the hard-rock gold mines, the Libby Creek Recreational Gold Panning Area is rich in placer gold deposits. Since the Pleistocene era, and until about 10,000 years ago, glaciers that once covered the area eroded away the edges of mountains and carried rock debris into what later became the Libby Creek drainage area. Gold, along with other heavy minerals, became concentrated as the material was worked through the flowing waters, leaving behind the heavy metals.

Professional gold panner Clayton Engman works the Libby Creek placers.

Plenty of gold can still be found along Libby Creek today. The site is open to the public and free of charge. Any gold you find is yours to keep. Prospecting is limited to panning and hand tools. No motorized or mechanized mining is allowed. While panning for the ancient gold in the cool water, examine the Precambrian rock, now smooth from the long journey. You can almost imagine what the landscape must have been like when the ancient Belt rocks were being carved by glaciers.

If you catch a hint of "gold fever" while on Libby Creek, keep in mind that the Northwest Montana Gold Prospectors club also owns a claim on Libby Creek and, for a small fee, their extensive claims are also accessible. For information contact Northwest Montana Gold Prospectors (NWMGP), PO Box 3242, Columbia Falls, MT 59912; (406) 892-3722.

The mining that perhaps made Libby most famous was the extensive deposits of hydrous biotite mica, called vermiculite, that were mined several miles northeast of Libby from the 1920s to 1990. When active, the mine supplied 80 percent of the vermiculite used worldwide. Vermiculite is a material that expands when heated. It has many uses; its primary use was as an insulation material. Almost ten years after the mine was out of operation, the Environmental Protection Agency determined that the vermiculate was a host of airborne cancer-causing asbestos and responsible for asbestos-related disease and fatalities. Today the Environmental Protection Agency oversees the Libby Superfund cleanup site.

4. Fishtrap Creek Fossils

See map on page 26.
Land type: Mountains, roadcut
GPS: N47°49.38480' / W115°08.93700'
Best season: Summer
Land manager: USFS, Lolo National Forest
Material: Fossils
Tools: Rock hammer
Vehicle: High clearance
Accommodations: Camping and RV parking in Kootenai National Forest; camping, RV parking, and motels in Libby
Special attractions: The Heritage Museum in Libby has exhibits on mining, forestry, and local history.

Finding the site: Some of these forest roads are difficult to locate so a forest map is recommended. From Libby drive southeast on US 2 and turn right onto McKillop Road (FR 535) toward the Fishtrap Creek (not lake) campground. The road will change numbers from FR 535 to FR 4422 and then to FR 516; the important thing to remember is to follow the signs to Fishtrap Creek (not lake) campground. Drive a total of 22.5 miles, turn left onto Shale Ridge Road (FR 7691), and continue 0.7 mile to a hill on the right side of the road with very thin gray shale layers eroding. Look carefully—the shale is so thin and easily weathered that it looks like dirt.

Rockhounding

The paper-thin layers of Cambrian shale contain fossilized casts of trilobites, a marine arthropod characterized by a beetle-looking three-lobed ovoid outer skeleton. Trilobites are from the Lower Cambrian (beginning of the Cambrian time), and their existence stretched to the Permian period. Trilobites existed for about 300 million years and thousands of species evolved, but with very little variation. The Libby-area trilobite fossils vary from about ¼ to ⅜ of an inch in size and less commonly up to 1½ inches in size. Most trilobite fossils are actually fossils of the shed skeletons and furthermore a cast, or internal mold, of the shed skeleton. On rare occasions the complete trilobite can be found, and on even rarer occasions a fossil of the underbelly of the trilobite is found. Nevertheless, these little critters are excellent additions to fossil collections. Brachiopods are less common than the trilobites but can be found at this site as well.

The shale emerging from the roadcuts is difficult to spot. Like many brittle shales, it erodes at a rate so extreme that it appears as dirt. Layers containing fossils can be exposed by brushing away shale debris and exposing hard layers of shale. The shale is still extremely brittle and easily turns to mud during rain. The layers have to be carefully split to expose trilobites and brachiopods and, due to their delicacy, diligent searching is required to collect a complete sample. Bring packing material and a secure container to transport delicate specimens.

5. Cox Gulch Minerals

See map on page 26.
Land type: Mountains
GPS: N47° 32.775' / W115° 35.603'
Best season: Spring through fall
Land manager: Private, United States Antimony Corporation
Material: Stibnite, pyrrhotite, galena, and sphalerite
Tools: Rock hammer
Vehicle: Any
Accommodations: Camping and RV parking nearby in Lolo National Forest; camping, RV parking, and motels in Thompson Falls
Special attractions: The restaurant, lounge, and lobby of the Rimrock Motel in Thompson Falls has a mining-theme decor, with several glass cases of minerals on display and available for purchase.

Finding the site: From Thompson Falls drive west on MT 200 and turn left onto Prospect Creek Road, just west of the Clark Fork River on the outskirts of town. Drive 13.5 miles and turn right at Cox Gulch and drive to the parking area. Inquire at the United States Antimony Corporation (USAC) office located in the parking area for access. It is recommended that you write in advance for permission to check tailings, or call ahead to ensure your arrival during the office hours. For more information contact the United States Antimony Corporation, PO Box 643, 1250 Prospect Creek Rd., Thompson Falls, MT 59873; (406) 827-3523.

Rockhounding

Though numerous mining districts can be found within Sanders County, those in the immediate vicinity of Thompson Falls are currently the most important. The earliest mining probably began around here in the 1880s, during the Coeur d'Alene gold rush. Little is known concerning production then, but it is certain that fair quantities of gold were recovered from placer deposits. Today the Prospect Creek District southwest of Thompson Falls is of particular importance because of the presence of antimony-bearing quartz veins within the Precambrian rocks. Mines are being developed by the US Antimony Corporation and are conveniently easy to access through the USAC office. Permission to dig through the tailings must be requested either in person or by mail, and at the time of my visit, it was nicely granted. Someone in the office will recommend directions to the tailings piles to ensure no disturbance on active areas. At the time of my visit, a dump pile of low-grade ore produced nice samples of stibnite, the principle ore of antimony, and stibnite crystals. The stibnite has a bright metallic silver luster, and the prismatic orthorhombic crystals are small but worth the search. With some luck, micromount quality of other minerals such as pyrrhotite, galena, and sphalerite may be found.

6. Taft Mountain Pyrite

See map on page 26.
Land type: Mountains
GPS: N47º 28.294' / W115º 33.892'
Best season: Summer
Land manager: USFS, Lolo National Forest
Material: Quartz, pyrite, dendrites
Tools: None
Vehicle: High clearance
Accommodations: Camping and RV parking nearby in Lolo National Forest; camping, RV parking, and motels in Thompson Falls and Superior
Special attractions: The restaurant, lounge, and lobby of the Rimrock Motel in Thompson Falls has a mining-theme decor, with several glass cases of minerals on display and available for purchase.

Finding the site: From the rest area on I-90 about 5 miles east of the Idaho Border, take exit 5 to the frontage road on the south side of the interstate. Drive east about 1 mile west of Saltese and 5 miles east of the Idaho border, head east for 2 miles, and left to head north on Randolph Creek Rd (FR 286). Drive north 2.9 miles and turn right onto FR 7709. Continue north up the steep grade for 4.9 miles to the intersection of FR 266 and FR 16996. Park and explore the area and east along FR 16996 to the top of the peak. You can also continue to the north side of the mountain instead for beautifully banded sandstones.

Rockhounding

Pyrite cubes can be easily collected in the area around Taft Mountain. Here you can search the loose pieces of quartz on the peak for pyrite cubes of various sizes. Also common to the area are dendrites. Often mistaken for fossils, dendrites are magnesium growths in plume patterning along rocks. On the north side of the mountain, you can find bright-colored banded sandstones.

Site 7

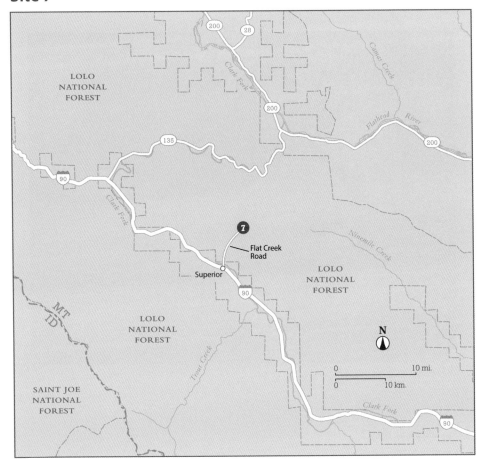

7. Superior Minerals

The tailings piles of Superior

See map on page 33.
Land type: Mountains
GPS: N47°14.43000' / W114° 51.30540'
Best season: Spring through fall
Land manager: USFS, Lolo National Forest
Material: Pyrite, arsenopyrite, quartz
Tools: None
Vehicle: Any
Accommodations: Camping and RV parking nearby in Lolo National Forest; camping, RV parking, and motels in Superior
Special attractions: The Mineral County Museum in Superior has displays on mining and local history.

Finding the site: From I-90 in Superior, take exit 47 north about 0.5 mile and turn left onto Mullan Road. From Mullan Road turn north onto Flat Creek Road. Drive 2 miles to a fork in the road, at which point you stay left and continue 1.5 miles to large piles of mine tailings on the left side of the road.

Rockhounding

Considerable quantities of zinc, lead, copper, silver, and some gold have been recovered from the Superior/St. Regis area of Montana during the past one hundred years. The ores were found in veins associated with several major fault zones. The majority of the rocks occurring here are metamorphic types of the Precambrian Belt series. Some younger igneous intrusions also occur throughout the area. Many of the mineral deposits that have been mined can be traced into the Coeur d'Alene District in Idaho just to the west.

The samples of pyrite and arsenopyrite at Superior are mediocre for collection due to weathering. I didn't spend much time at this site, and there may be more minerals waiting to be discovered. Most of the tailings piles do have nice pieces of quartz large enough to cut or sphere. Tailings occur throughout the area, so don't feel limited to exploring just the suggested locality. Farther up the road from the site mentioned, there are several more piles, along with a warning about a mysteriously bright-orange arsenic-bearing "toxic creek."

Of course, it is recommended that you stay out of the creek. Like many old mining sites, the charm and history come along with a significant cost to the environment.

Sites 8 and 9

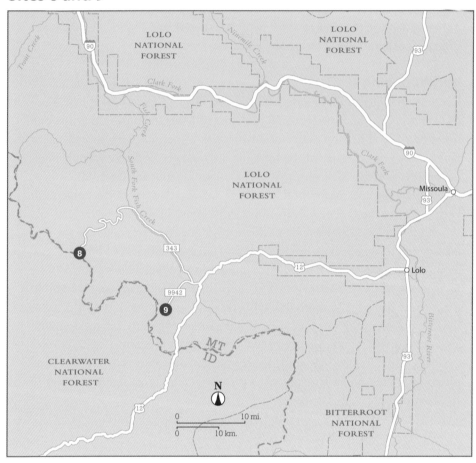

8. Snowbird Mine Fluorite

See map on page 36.

Land type: Mountains

GPS: N46° 46.59780' / W114° 47.63340'

Best season: Summer

Land manager: USFS, Lolo National Forest

Material: Quartz crystals, fluorite, parisite

Tools: Rock hammer

Vehicle: Any

Accommodations: Camping and RV parking in Lolo National Forest; camping, RV parking, and lodge in Lolo and resort town of Lolo Hot Springs

Special attractions: Lolo Hot Springs is a natural source of geothermal water. It has been developed into a resort and provides weary vacationers with relaxation and comfortably warm water in a well-kept pool.

Finding the site: From the junction of US 93/12 in Lolo, drive west on US 12 for 25.5 miles. Just past the Lolo Hot Springs resort area, turn right onto Fish Creek Road (FR 343). Take Fish Creek Road for 13.8 miles and turn left onto the road leading toward the Schley Mountain Trailhead. Drive 11.7 miles, stay left at the fork, and drive toward the upper trailhead, located 2.9 miles past the fork. At the time of my visit, the road that leads to the mine from this point was gated off and closed to motorized traffic past the trailhead. You may have to continue up this road by foot; park at the trailhead and walk about 1 mile farther on the road, not on the hiking trail. It is an uphill hike to the mine once you pass the gate. When the old mining road splits about halfway to the mine, stay on the low (right) path. Many of the forest roads leading to Snowbird Mine do not open until mid- to late summer, so always inquire locally about accessibility.

Rockhounding

About 77 million years ago, a carbonatite sill intruded along a fault in sedimentary Belt series rocks. The Snowbird Mine developed the fluorite associated with the fault's sill. The large deposits appear to have been mined for quite some time, but today the mine is long since defunct. For the rock hound this abandoned mine is a treasure trove. The old road to the mine sparkles with scattered fluorite like a trail of sprinkled diamonds. Once at the mine, high up in the mountain, the steep hike is rewarded with a spectacular view of some of Montana's most lush territory. Veins of fluorite were mined by blasting the mountainside, and several large caves exist as scars to the story. The fluorite veins inside the caves are bright shades of green, white, and a deep blackish purple. Even though acres of dumps are located beneath the mine, some of the best samples can be collected from the source in the caves and on the mountainside. Parisite crystals up to 9 inches long have been found at the Snowbird Mine.

Large chunks of quartz blasted from the mountain rest in piles along the edges of the caves. Translucent quality quartz crystals occur within some of these rocks up to several inches long, but it's unlikely they can be removed without damaging the crystal. Low-grade quartz crystals without any translucency occur up to a weight of a couple hundred pounds. These behemoths are often up to a foot in diameter, and they're impossible to remove without damage to the crystal. The beautiful fairy-tale size of the crystals is best viewed and left untouched for other rock hounds to enjoy. Any attempt at removing the crystals would only result in damaging their beauty and most likely your tools.

Veins of fluorite and Snowbird Mine

Exploring the fluorite caves at
Snowbird Mine

9. Lolo Crystals

See map on page 36.
Land type: Mountains, roadcut
GPS: N46° 41.89080' / W114° 36.28080'
Best season: Late spring through fall
Land manager: USFS, Lolo National Forest
Material: Quartz crystals
Tools: None
Vehicle: Any
Accommodations: Camping and RV parking in Lolo National Forest; camping, RV parking, and motels in Lolo and the resort town of Lolo Hot Springs
Special attractions: Lolo Hot Springs is a natural source of geothermal water. It has been developed into a resort and provides weary vacationers with relaxation and warm water.

Finding the site: From the junction of US 93/12 in Lolo, drive west on US 12 for about 25 miles. Go 1 mile past the Lolo Hot Springs resort area and turn right onto FR 343. Drive 2 miles along FR 343 and turn left onto Granite Creek Road. Drive 5 miles down to the site; look for evidence of digging in the roadcuts.

Rockhounding

In the area north and west of Lolo Hot Springs, along Granite Creek and its tributaries, numerous outcrops of the granitic Idaho batholith are accessible because of extensive logging. Beautiful crystals of clear to very dark smoky quartz can be found in miarolitic cavities within the rock and even loose in the soil. Their occurrence and digging for them in the decomposed granitic soil is somewhat similar to Crystal Park.

The directions mentioned will take you to a general location where crystals can be found. The most popular and productive ridge along Granite Pass has been closed due to excessive digging. Due to the gaping wounds of the forest, collecting is no longer permitted in this area.

Most of the crystals in this area are found loose in the soil on the surface where the granite is decomposing. The directions to this location are outside the closure zone. Digging is not permitted but surficial collecting is. There is still plenty of evidence where others continue to dig anyway and this is not advised or necessary. In order to locate other areas where crystals may occur, especially in hard rock, carefully examine the rock exposures in roadcuts, hillsides, and exposed areas of recent logging. In particular, look for patches of smoky quartz or the presence of unusually large grains of quartz or feldspar (pegmatite) within the smaller grained granite. Most of the rock in the area is generally uniform in texture, with relatively small grains of quartz and feldspar, and so the pegmatite nature of a rock outcrop may indicate a potential source of well-formed crystals. The area of granite to be explored spans for miles and is in no apparent structure. The crystals occur throughout the forest.

Extreme care should be taken in extracting the crystals. Many beautiful specimens have been destroyed by hasty decisions and poor judgment in removing them from the host rock. Groups of brilliant crystals of various sizes, ranging from pounds heavy to less than an inch big, can be the prize of a patient collector. The crystals uncovered from the loose soil may be smaller, usually less than an inch in size, but the work is also easier and gem quality can still be found.

Sites 10–14

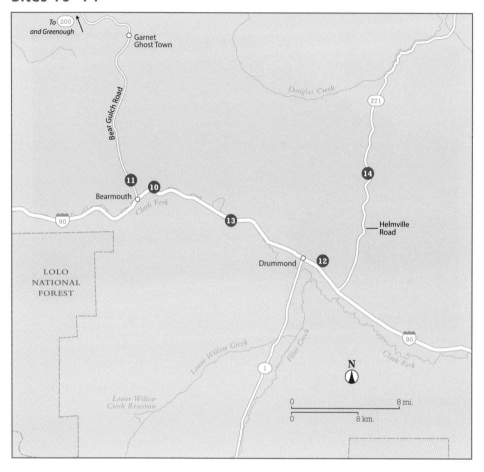

To ⬭200
and Greenough

Garnet
Ghost Town

Douglas Creek

271

Bear Gulch Road

11   10

Bearmouth

Clark Fork

90

13

Helmville
Road

14

LOLO
NATIONAL
FOREST

Drummond   12

Lower Willow Creek

Flint Creek

Clark Fork

90

1

Lower Willow
Creek Reservoir

N

0   8 mi.
0   8 km.

10. Garnet Mountain Fossils

Limestone outcrops of the Garnet Mountains

See map on page 41.
Land type: Mountains, roadcut
GPS: N46° 43.154' / W113° 18.746'
Best season: Spring through fall
Land manager: Bureau of Land Management
Material: Fossils, calcite crystals, chert
Tools: None
Vehicle: Any
Accommodations: Camping and RV parking in Lolo National Forest; camping, RV parking, and motels in Deer Lodge and Missoula
Special attractions: Garnet ghost town is located up Bear Gulch about 10 miles north of the I-90 frontage road. For more information visit garnetghosttown.net.

Finding the site: From Drummond on I-90 take exit 153 to the north side of the freeway. Drive northwest on the frontage road 9.4 miles to the interpretive sign about Madison Limestone on your right.

Marine fossils from the Garnet Mountains

Rockhounding

This scenic site is located in the Mississippian Madison Limestone, a formation laid down about 350 million years ago. An excellent Montana Department of Transportation roadside marker describes the more than 1,000-foot-thick formation and the burial setting of the many marine critters to be found. Begin your hunt at the sign and in the gully and hill to the north where you will find bivalve, crinoid, and other marine organisms that flourished in the warm tropical ocean shelf that later hardened into the limestone.

Very nice samples of calcite crystals can also be found along the road and hillside. Samples of small clusters of crystals are already exposed and can be easily collected. Most specimens of the attractive clusters can't be removed from the rock without damaging it, but they are still quite nice. Drusy crystal blankets and chert nodules are also abundant.

11. Garnet Minerals

See map on page 41.

Land type: Mountains, tailings

GPS: N46° 43.501' / W113° 20.277'

Best season: Spring through fall

Land manager: USFS, Lolo National Forest

Material: Quartz, calcite, pyrite, fossils, arsenopyrite, garnet, chert

Tools: None

Vehicle: Any

Accommodations: Camping and RV parking in Lolo National Forest; camping, RV parking, and motels in Deer Lodge and Missoula

Special attractions: Garnet ghost town is located up Bear Gulch about 10 miles from this site. For more information visit garnetghosttown.net.

Finding the site: From Drummond on I-90 take exit 153 to the north side of the freeway. Drive northwest on the frontage road 10 miles and turn right (north) onto Bear Gulch Road. The tailings begin about 0.5 mile up the road. There is a mixture of USFS and private land along the tailings so watch for No Trespassing signs and respect fences. To get to Garnet, head 7.5 miles up Bear Gulch Road to Cave Gulch Road, following the signs for 4 more miles to Garnet. Note that collecting is not permitted within the protected ghost town and should be kept well outside the town.

Rockhounding

Garnet is the remnants of a late-1800s mining town named after the mineral itself. After the discovery of gold, miners flocked to the region and the town hosted four hotels and thirteen bars. Today miles of tailings piles and decaying buildings are all that is left. The ghost town is a popular tourist attraction, and collecting is not permitted within the protected area. The tailings piles and other signs of mining line Bear Gulch for several miles, and there are plenty of collecting areas within the general USFS land. The tailings are an interesting myriad of rock and mineral types, all formerly the gold-rich placers of a decimated creek. Rocks range from local limestones and granites to well-rounded garnet schists and quartz cobbles. Nothing too spectacular was found here, but a trip to the region is not complete without visiting the ghost town so it's worth a stop.

12. Drummond Fossils

Drummond fossils and calcite

See map on page 41.

Land type: Mountains, roadcut

GPS: N46° 40.095' / W113° 07.875'

Best season: Spring through fall

Land manager: Powell County

Material: Fossils, calcite crystals

Tools: None

Vehicle: Any

Accommodations: Camping and RV parking in Lolo National Forest; camping, RV parking, and motels in Drummond, Deer Lodge, and Missoula

Special attractions: Garnet ghost town is located up Bear Gulch about 10 miles north of the I-90 frontage road. For more information visit garnetghosttown.net.

Finding the site: From Front Street in Drummond, head east to the edge of town, turn north onto the unlabeled road (Sorenson Lane), and drive 0.3 mile to the large hill on the left just past the landfill.

Rockhounding

Fossiliferous Cretaceous sandstones and limestones surround the Drummond vicinity. At this particular location, limestone of the lower Cretaceous Kootenai Formation is well exposed and highly fossiliferous. Collecting is very easy and requires no tools. If you face the outcrop and hill, the gray limestone is best exposed to the left side of the cut. Large pieces of limestone are loose and contain abundant marine fossils, mainly of gastropods (snails) and bivalves. You can also find large chunks of opaque calcite displaying the rhombohedral nature of calcite and smaller crystal pockets in the limestone with more translucent, intergrown crystals and drusy. There is a lot of material to be found at this location, but the smell of the nearby landfill will likely limit the length of your visit.

13. Rattler Gulch Fossils

See map on page 41.
Land type: Hills, roadcut
GPS: N46° 41.789' / W113° 13.802'
Best season: Spring through fall
Land manager: Montana DOT
Material: Fossils
Tools: Rock hammer
Vehicle: Any
Accommodations: Camping and RV parking in Lolo National Forest; camping, RV parking, and motels in Deer Lodge and Missoula
Special attractions: Garnet ghost town is located up Bear Gulch about 10 miles north of the I-90 frontage road. For more information visit garnetghosttown.net.

Finding the site: From Drummond on I-90 take exit 153 to the north side of the freeway. Drive 4 miles northwest on the frontage road and turn left at the sign for Rattler Gulch. Drive a short distance, about 0.1 mile, through a very "colorful" tunnel under the freeway that was "decorated" by local graffiti artists. Turn right onto the gravel road that follows the railroad tracks and drive 0.2 mile to vertical layers protruding from a hill on the right.

Rattler Gulch fossils

Rockhounding

These steeply tilted rock layers are sedimentary deposits of the Jurassic and Cretaceous ages. The marine layers of the Jurassic rocks have produced some fairly well-preserved fossils. Several species of mollusks have been reported from the shales and limestones present near the east end of the railroad cut. Bivalves and the oyster Gryphaea are some of the more common fossils to be found here. The rock breaks easily, and weathered specimens can be found already broken free from the outcrop.

14. Helmville Fossils

A small cluster of snails and delicate leaf imprint at Helmville

See map on page 41.
Land type: Hills, roadcut
GPS: N46° 43.790' / W113° 05.018'
Best season: Summer through fall
Land manager: Montana DOT
Material: Fossils
Tools: Rock hammer
Vehicle: Any
Accommodations: Camping and RV parking in Lolo National Forest; camping, RV parking, and motels in Deer Lodge and Missoula
Special attractions: Garnet ghost town is located up Bear Gulch about 10 miles north of the I-90 frontage road. For more information visit garnetghosttown.net.

Finding the site: From I-90 in Drummond, take exit 154 to the south side of the highway. Drive southeast on the frontage road for 1.5 miles and turn left onto the road to Helmville (MT 271). Travel 6.5 miles to a large roadcut on the right side of the road at mile marker 8. The site is difficult to find since the rock containing the fossils is hard to see. Keep an eye out for the thin light-colored shale eroding from several hills facing the road.

Rockhounding

Extremely well-preserved fossilized leaves can be found in mid-Paleogene rocks northeast of Drummond. Gastropods (snails), insects, and a few rare fish skeletons have been reported in the area as well. The light-colored shale on the right side of the road contains the fossils. Nice specimens are not overly abundant, so you will have to look hard to find anything. Diligence can pay off, however, because the fossils tend to be finely detailed. Be sure to bring sufficient packaging materials to wrap anything collected. The brittle shale breaks easily.

Sites 15–17

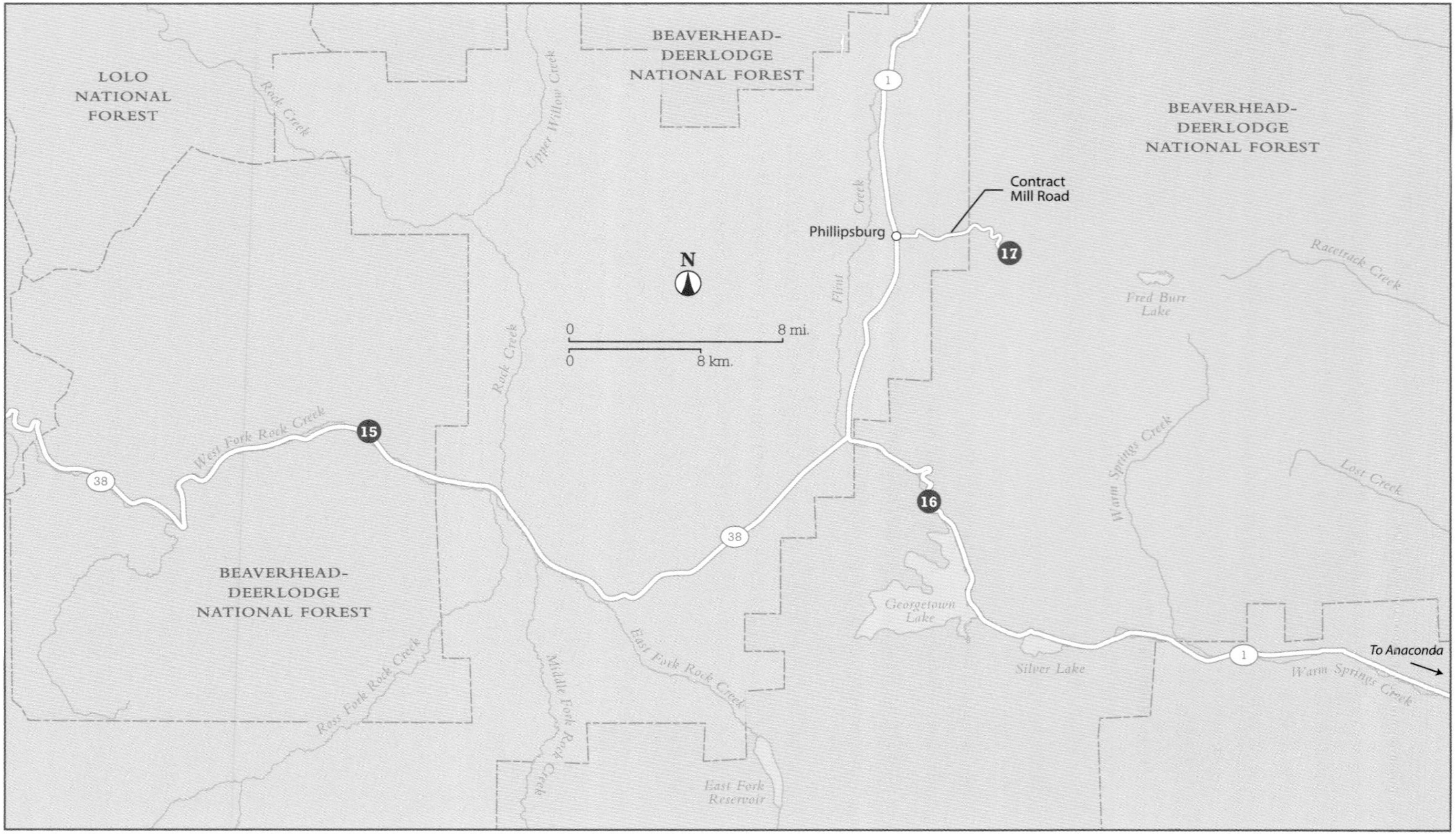

LOLO NATIONAL FOREST

BEAVERHEAD-DEERLODGE NATIONAL FOREST

BEAVERHEAD-DEERLODGE NATIONAL FOREST

BEAVERHEAD-DEERLODGE NATIONAL FOREST

Phillipsburg

Contract Mill Road

To Anaconda

Georgetown Lake

Silver Lake

Fred Burr Lake

Warm Spring Creek

Rantrack Creek

Lost Creek

Warm Spring Creek

Flint Creek

Upper Willow Creek

Rock Creek

Rock Creek

West Fork Rock Creek

East Fork Rock Creek

Middle Fork Rock Creek

Ross Fork Rock Creek

East Fork Reservoir

N

8 mi

8 km.

1

1

38

38

15

16

17

15. Gem Mountain Sapphires

See map on page 50.

Land type: Mountains

GPS: N46° 14.82420' / W113° 35.54100'

Best season: Generally open 7 days a week mid-May to Oct

Land manager: Private, Gem Mountain Cooney's Sapphire Village, Inc.

Material: Sapphires, garnet, ruby, quartz, feldspar, moonstone, jasper

Tools: Supplied by business

Vehicle: Any

Accommodations: Camping and RV parking nearby in Beaverhead-Deerlodge National Forest; camping, RV parking, and motels in Philipsburg and Anaconda

Special attractions: The Granite County Museum and Cultural Center in Philipsburg has displays on mining and Montana history, including a sample of a Philipsburg raw ore vein. In Anaconda at the historic district, you can view a 585-foot-tall smokestack—all that remains of "Copper King" Marcus Daly's 19th-century smelter. About 4 miles east of Philipsburg are the remains of the ghost town of Granite, one of the more colorful mining camps of Montana's past.

Finding the site: Philipsburg is located between Glacier National Park and Yellowstone National Park along the Pintler Scenic Route, a loop connecting to I-90 on both ends. From Philipsburg drive south on MT 1 for 6 miles to the junction with MT 38. Take MT 38 (Skalkaho Pass) 16 miles west to Gem Mountain. There are signs for Gem Mountain posted on MT 38, and a large welcome sign is at the driveway to the mine.

Rockhounding

The sapphire deposits at the Gem Mountain locality are somewhat similar to those in the Helena area, but they were not deposited by a stream the size of the Missouri River. These deposits include bench and terrace gravels that were laid down by mountain streams.

Mining began at Gem Mountain in the early 1890s, and since that time it has produced more than 180 million carats in sapphires. It has been reported that the sapphires found at Gem Mountain exhibit a wider range of colors (from yellow to pink and lavender) than those found at other localities in the state.

There is no digging directly at Gem Mountain, but customers may purchase buckets of pre-dug gravel from the mine for screening on the premises. All of the tools needed are provided on-site, along with a staff to help you screen your purchases. Bags of prescreened "pay dirt" concentrate are available for purchase, and bags can be shipped worldwide if customers are interested in looking for sapphires in the comfort of their own homes. The owners of the mine also provide a faceting service, and trained jewelers are on-site, along with a gift shop with a good selection of fine jewelry for purchase. If you don't make it out to the mine, you can also visit their larger jewelry store in town, which offers year-round sapphire sifting (gravel supplied per bags). For more information or to order gravel, contact Gem Mountain at (866) 459-GEMS or www.gemmountainmt.com.

Sifting for sapphires at Gem Mountain

Sapphires from a day at Gem Mountain

16. Flint Creek Rocks

Mudstone cliffs with mudcracks, ripple marks, and raindrop impressions

See map on page 50.

Land type: Mountains, roadcut

GPS: N46° 13.25000' / W113° 17.28333'

Best season: Spring through fall

Land manager: Montana DOT

Material: Mudcracks, ripple marks, raindrop imprints

Tools: Rock hammer

Vehicle: Any

Accommodations: Camping and RV parking in Beaverhead-Deerlodge National Forest and on Georgetown Lake; camping, RV parking, and motels in Philipsburg and Anaconda

Special attractions: The Granite County Museum and Cultural Center in Philipsburg has displays on mining and Montana history, including a sample of a Philipsburg raw ore vein. In Anaconda at the historic district, you can view a 585-foot-tall smokestack—all that remains of "Copper King" Marcus Daly's 19th-century smelter. About 4 miles east of Philipsburg are the remains of the ghost town of Granite, one of the more colorful mining camps of Montana's past.

Mudcracks at Flint Creek Hill

Finding the site: Philipsburg is located between Glacier National Park and Yellowstone National Park along the Pintler Scenic Route, a loop connecting to I-90 on both ends. From Philipsburg drive south on MT 1 for about 10 miles to Flint Creek Hill, a large roadcut between Georgetown Lake and the Philipsburg Valley. Towering vertically lifted Precambrian mudstones are on both sides of the road. Even the boulder-size slabs of Precambrian rock placed by the highway department in the turnout to keep cars from plunging into Flint Creek hold specimens of mudcracks that weather away from the rock in nicely split layers.

Rockhounding

This is by far the best location in Montana to get in touch with Precambrian time. The Precambrian sedimentary layers of the Belt series are at a vertical stance and tower above the highway at a minimum of 30 feet high. In these layers you can see the evidence of a time long since gone on earth. The billion-year-old Precambrian stone contains well-preserved mudcracks and ripple marks and reports of raindrop impressions. This rock started as the sediments of a shallow sea that once covered the land on earth before the atmosphere was rich in oxygen and before animals existed. The imprints in the mudstone technically aren't considered fossils, because they leave no evidence of life. They serve as evidence of a time before life, a time before animal tracks or even burrowing casts from worms. The layers of imprints separate easily, and tools are not required because the road work has flaked off good pieces.

17. Philipsburg Minerals

See map on page 50.
Land type: Mountains, tailings
GPS: N46° 19.029' / W113° 14.675'
Best season: Spring through fall
Land manager: USFS, Beaverhead-Deerlodge National Forest
Material: Rhodochrosite, galena, silver, sphalerite, pyrite, pyrolusite, arsenopyrite, tetrahedrite
Tools: None
Vehicle: Any
Accommodations: Camping and RV parking in Beaverhead-Deerlodge National Forest; camping, RV parking, and motels in Philipsburg and Anaconda
Special attractions: The Granite County Museum and Cultural Center in Philipsburg has displays on mining and Montana history, including a sample of a Philipsburg raw ore vein. In Anaconda at the historic district, you can view a 585-foot-tall smokestack—all that remains of "Copper King" Marcus Daly's 19th-century smelter. About 4 miles east of Philipsburg are the remains of the ghost town of Granite, one of the more colorful mining camps of Montana's past.

Finding the site: The city of Philipsburg is located between Glacier National Park and Yellowstone National Park along the Pintler Scenic Route, a loop connecting to I-90 on both ends. Philipsburg publishes a free newspaper guide for tourists that is available throughout the town. The *Philipsburg Territory* is updated every year and includes a historic driving tour of Philipsburg and the mining district just outside of town. The tour is printed in the paper along with a nice map. Each mine is identified on the map, and the land manager is noted. Be aware that public and private land is intermingled throughout the district. It is a good idea to pick up one of these maps since many of the mines are on private property. Always be on the lookout for any indication of private property on the dozens of abandoned mines east of Philipsburg. A good starting point is Granite ghost town. From

The old mines of Granite ghost town

downtown Phillipsburg head south on Sansome Street/Highway 10 for 0.5 mile and turn left onto Granite Street. Drive 3.5 miles on Granite Street following the signs to the old mines and ghost town. For a copy of the *Philipsburg Territory* or more information, contact Philipsburg Chamber of Commerce, at (406) 859-3388; www.philipsburgmt.com.

Rockhounding

The Philipsburg District was a relatively important one during the early 1900s, after the discovery of gold and silver near Georgetown to the south. Several deposits have been exploited in the district, and the presence of a smelter indicates an extensive source of ore.

The occurrence of such minerals as pyrolusite, psilomelane, and rhodochrosite attest to the fact that manganese was one of the major elements being searched for here. Fragments of the host rock are common on the dumps and tailing, and cavities or vugs in this rock may reveal outstanding tiny crystals of manganese minerals. The possibility of obtaining some interesting micromount specimens in this area is good.

The best luck in collecting good samples from this large district may be to use the *Philipsburg Territory* map of the mining district and spend a day driving throughout the district and exploring some of the dozens of tailings piles in the area. Granite ghost town is a good starting point. There are a lot of interesting minerals to be found here, so don't forget to bring a rock-and-mineral-identification guide—who knows what you may find?

Sites 18–20

18. Blackfoot River Rocks

See map on page 57.

Land type: River, gravel bars

GPS: N46° 56.289' / W112° 52.618'

Best season: Spring through fall

Land manager: USFS, Helena National Forest

Material: Agate, jasper, chalcedony

Tools: None

Vehicle: Any

Accommodations: Primitive camping on-site, camping and RV parking in Helena National Forest; camping, RV parking, and motels in Lincoln

Special attractions: None

Finding the site: From Lincoln drive 12.3 miles west on MT 200 to BLM signs for the Blackfoot River fishing access on the left; the parking area is on the right. Enter the public fishing area and follow the trails to gravel bars and clearings in the plateaus along the river. Public access signs are very helpful in identifying the area open for public use. Another access point is the unmaintained Nevada Ogden campground along the Blackfoot River 2 miles east of the fishing access.

Rockhounding

The Blackfoot River doesn't host the same geologic deposits as the Yellowstone River and the agate is not nearly as prized. Along the Blackfoot River you can find pebble-to-cobble-size agates, chalcedony, and jaspers. The agate is not as translucent as the Yellowstone and also lacks moss agate. The access to this fishing area is behind a barbed-wire fence, and in typical rugged Montana fashion, BLM had built a steep double-sided ladder over the fence to allow public access instead of installing a gate. Once over the ladder, the agate is found in the gravel clearings, but it is not abundant. The agate is usually red, brown, gray, and green. You can also find other samples of chalcedony and jasper with nice patterning.

19. Rogers Pass Onyx

See map on page 57.

Land type: Mountain pass, roadcut

GPS: N47° 04.830' / W112° 22.163'

Rogers Pass onyx

Best season: Summer

Land manager: USFS, Helena National Forest

Material: Onyx, chalcedony, quartz

Tools: Rock hammer

Vehicle: Any

Accommodations: Camping and RV parking in Helena National Forest; camping, RV parking, and motels in Lincoln

Special attractions: None

Finding the site: From Lincoln drive 18.5 miles east on MT 200 and turn right onto a small unlabeled gravel road immediately after the pass and before the guardrail on the right (Little Wolf Creek Road). Drive 0.1 mile to the outcrop on the right. You can also walk from the pass parking area if your car is not equipped for the unmaintained road.

Rockhounding

This unique deposit of silica-rich precipitant consists mainly of chalcedony. The roadcut is located at a high elevation just above Rogers Pass near Lincoln. Chalcedony is a variety of quartz that includes agate, jasper, and onyx. Onyx refers to the variety of chalcedony that is banded. Here the banding occurs in shades of whites and yellow. The rock is not easily broken free from the road-cut so it is best to search for loose pieces. You can also find pockets of quartz crystal drusy and solid small veins of milky white common opal.

20. Bowman's Corner Fossils

See map on page 57.

Land type: Hills, roadcut

GPS: N47° 15.197' / W112° 13.200'

Best season: Spring through fall

Land manager: Montana DOT

Material: Fossils, petrified wood, selenite, gypsum

Tools: None

Vehicle: Any

Accommodations: Camping and RV parking in Helena National Forest; camping, RV parking, and motels in Lincoln
Special attractions: None

Finding the site: From the junction of MT 200 and US 287 in Bowman's Corner, head west and explore the roadcuts from 2 to 4 miles. The particular location is 3.8 miles west on MT 200 on the south side of the road between mile markers 105 and 106.

Rockhounding

The Two Medicine Formation is one of the more important dinosaur-bearing formations in the world and contributed to the fame of Montana and its dinosaurs. The formation was not heavily studied due to poor exposures. Although it was described and mapped in the 1930s, it wasn't until forty years later when Marion Brandvold, a local rock hound who ran a small rock shop in Bynum, discovered baby dinosaurs that the formation became famous. Later studied and published on, Bynum's Egg Mountain consisted of hundreds of skeletons of hadrosaur (duckbill) dinosaurs and eggs. The large amount of eggs and baby skeletons combined with the adult skeletons at Egg Mountain heavily influenced the idea that dinosaurs may have been less like cold-blooded lizards who simply lay their eggs and move on and more like warm-blooded mammals who rear their young.

At this site near Bowman's Corner you can explore the Two Medicine Formation and collect plant fossils, petrified wood, and gypsum. A nice variety of selenite can also be found. The transparent, chunky crystal, which name stems from the Greek word meaning "moon." No vertebrate fossils or eggs were located at this site, and keep in mind that it is illegal to collect them. If you are interested in joining a dinosaur dig in the Two Medicine Formation, see Site 23 for the Two Medicine Dinosaur Center, where Marion's son, degreed paleontologist Dave Trexler, will organize you on a dig.

Sites 21 and 22

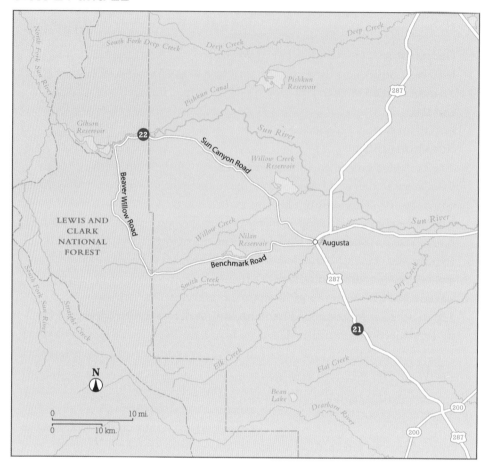

LEWIS AND CLARK NATIONAL FOREST

Beaver Willow Road

Sun Canyon Road

Benchmark Road

Augusta

N

0   10 mi.
0   10 km.

21. Augusta Fossils

See map on page 62.
Land type: Mountains, roadcut
GPS: N47° 23.34900' / W112° 18.92520'
Best season: Spring through fall
Land manager: Montana DOT
Material: Fossils
Tools: None
Vehicle: Any
Accommodations: Camping and RV parking in Lewis and Clark National Forest by Augusta; camping, RV parking, and motels in Great Falls
Special attractions: The Two Medicine Dinosaur Center in Bynum has a museum with excellent dinosaur and geologic displays. The Old Trail Museum in Choteau has exhibits on local history and paleontology. The Sun River Canyon, located 15 miles north of Augusta on Sun River Road, is a spectacular canyon carved into Madison limestone.

Finding the site: From Bowman's Corner at the junction of MT 200 and US 287, drive north on US 287 for 10.2 miles to a roadcut on the right. The most fossiliferous layers are located on the closest side of the roadcut to Bowman's Corner, and there is no need to pass the bend. It is best to rockhound on this side of the cut; otherwise, a blind spot is created for oncoming traffic. Exercise extreme caution when anywhere around speeding traffic.

Rockhounding

In the area near Augusta, a thick concentrated layer of solid oyster fossils in sandstone can be found. The Cretaceous oysters are around 80 million years old and still retain their mother-of-pearl shell (nacre), but they are almost impossible to break from the roadcut. The best luck collecting good specimens is found by searching for fossils in loose pieces that have already fallen free.

22. Sun River Fossils

See map on page 62.
Land type: Mountains, roadcut
GPS: N47° 37.12140' / W112° 42.09660'
Best season: Spring through fall
Land manager: Montana Fish, Wildlife, and Parks; Sun River WMA
Material: Fossils, calcite crystals
Tools: None
Vehicle: Any
Accommodations: Camping and RV parking on-site and in Lewis and Clark National Forest; camping, RV parking, and motels in Great Falls and Choteau
Special attractions: The Two Medicine Dinosaur Center in Bynum has a museum with excellent dinosaur and geologic displays. The Old Trail Museum in Choteau has exhibits on local history and paleontology.

Finding the site: The rocks of the Sun River Canyon and vicinity are predominantly fossiliferous limestone, and this is only one recommended starting point. From Augusta at the junction of Main and Manix Streets, drive 3.5 miles west on Manix Street and turn right onto Sun Canyon Road toward Sun River Canyon. Explore the limestone on the drive in and around the canyon, about 15 miles down the road.

Rockhounding

Through a series of faults, the overthrust belt of the Sawtooth Mountains has placed Mississippian Madison limestone above younger Cretaceous shales and sandstones. The Sun River flows through a canyon, exposing the various ages of the strata. Nice specimens of horn coral and brachiopods can be found in the Madison limestone throughout the canyon area. The limestone is extremely difficult to break and the best fossil collecting may be found by exploring the loose pieces of rock along the base of the roadcut that has already weathered from the cliffs. Think of the particular patch mentioned above as a starting point.

Sites 23 and 24

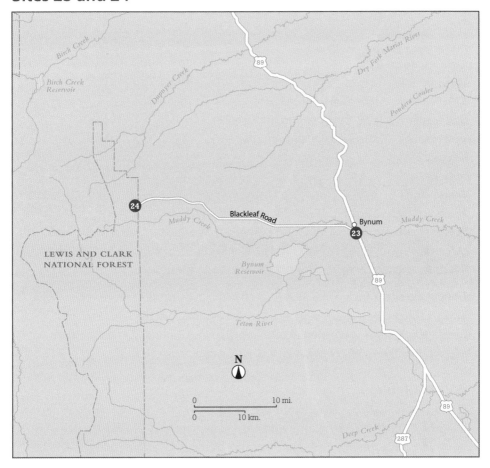

23. Two Medicine Dinosaur Digs

See map on page 65.
Land type: Hills
GPS: See contact information below
Best season: Generally operates May through Oct
Land manager: Private, Timescale Adventures
Material: Fossils
Tools: None
Vehicle: Any
Accommodations: Camping, RV parking; motels in Choteau
Special attractions: Old Trail Museum in Choteau has exhibits on local history and paleontology. Sun River Canyon, located 15 miles north of Augusta on Sun Canyon Road, is a spectacular canyon carved into Madison Limestone.

Finding the site: Digs are generally located in the Bynum/Choteau area of Montana along the Rocky Mountain front. The digs begin at the Two Medicine Dinosaur Center in Bynum. Contact the Two Medicine Dinosaur Center for more information: (406) 469-2211 or (800) 238-6873; www.timescale.org.

Rockhounding

If you can only join one dinosaur dig in Montana, this should be your destination. At the Two Medicine Dinosaur Center, you get the experience of a nonprofit dinosaur museum and digs run by degreed paleontologists along with the experience of digging with one of Montana's most famous dinosaur families. The Two Medicine Formation is one of the more important dinosaur-bearing formations in the world and contributed to the fame of Montana and its dinosaurs. The formation was not heavily studied due to poor exposures. Although it was described and mapped in the 1930s, it wasn't until forty years later when Marion Brandvold, a local rock hound who ran a small rock shop in Bynum, discovered baby dinosaurs that the formation became famous. Later studied and published on, Bynum's Egg Mountain consisted of hundreds of skeletons of hadrosaur (duckbill) dinosaurs and eggs. The large amount of eggs and baby skeletons combined with the adult skeletons at Egg Mountain heavily influenced the idea that dinosaurs may have been less like cold-blooded lizards who simply lay their eggs and move on, and more like warm-blooded mammals who protect and rear their young. The remarkable discovery also led to the dedication of the state fossil, the species of hadrosaur dinosaur of the mountain, *Maisaura peeblesorum*, meaning "good mother lizard."

The Two Medicine Dinosaur Center/Timescale Adventures is a nonprofit organization run by David Trexler, son of Marion Brandvold, who has spent his life dedicated to the dinosaurs of the region. The digs are based out of the Two Medicine Dinosaur Center in Bynum, located in the heart of the Two Medicine Formation. The center has a museum with extensive displays on paleontology and is home to a skeletal model of the world's longest dinosaur. Paleontology field trips into

Jacketing a dinosaur skeleton in plaster with Two Medicine Dinosaur Center

the Two Medicine Formation of the Choteau area with Timescale Adventures can range anywhere from a couple of hours to a ten-day course. A minimum of a daylong program is required to participate in an excavation. The courses are designed as "hands-on," and the staff incorporates an overview of geologic history, fossil identification, excavation, and preservation. Longer programs cover all aspects of field paleontology. Group sizes are generally limited and trips are tailored to the age and requests of the group. Reservations are recommended.

24. Blackleaf Fossils

See map on page 65.
Land type: Mountains
GPS: N 48° 055452' / W 112° 42.54072'
Best season: Summer
Land manager: Montana Fish, Wildlife, and Parks, Blackleaf WMA
Material: Fossils
Tools: None
Vehicle: Any
Accommodations: Camping on-site and in Lewis and Clark National Forest, RV parking, and motels in Choteau
Special attractions: The Two Medicine Dinosaur Center in Bynum has a museum with excellent dinosaur and geologic displays. The Old Trail Museum in Choteau has exhibits on local history and paleontology. Sun River Canyon, located 15 miles north of Augusta on Sun River Road, is a spectacular canyon carved into Madison Limestone.

Finding the site: From Bynum take 3rd Street west and continue as it turns into Blackleaf Road for 18 miles to the trailhead. Hike into the canyon and begin looking for fossils about 0.3 mile in.

Rockhounding
The Mississippian Madison Limestone is well exposed on this eastern flank of the Rocky Mountains. Here you can find a variety of marine fossils. The most common of the fossils are corals, bivalves, and crinoid stems.

Site 25

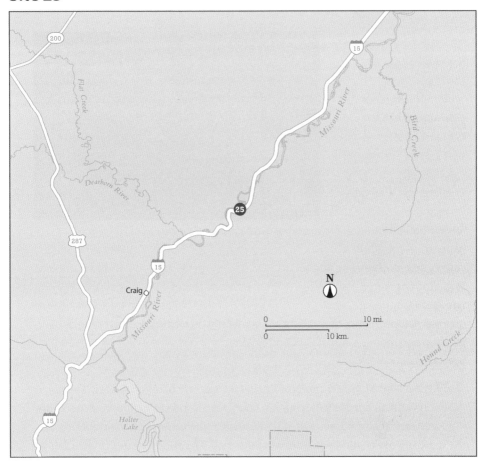

25. Craig Minerals

See map on page 69.
Land type: Low
mountains, roadcut
GPS: N47° 09.282' /
W111° 49.824'
Best season: Spring
through fall
Land manager: Cascade
County
Material: Stilbite,

laumontite, mesolite,
calcite, quartz, amethyst, augite
Tools: Rock hammer
Vehicle: Any
Accommodations: Camping and RV parking nearby on Holter Lake; camping, RV
parking, and motels in Great Falls and Helena
Special attractions: None

Augite crystals and zeolites of the volcanic rubble

Finding the site: From I-15 north of Craig, take exit 244 (Canyon Access). Drive
north on the frontage road (Recreation Road) for 0.5 mile to a large roadcut on the
right between the frontage road and I-15. Park in the turnout on the right side of
the road and use caution while crossing the road. The entire Recreation Road drive
is very scenic and highly recommended.

Rockhounding
Roadcuts between Craig and Hardy on the frontage roads to I-15 yield specimens of the zeolite minerals stilbite, laumontite, and mesolite, as well as calcite, rare amethyst, and augite crystals. Zeolite is a term used to describe any
number of minerals of the zeolite group, a group of hydrous aluminosilicates
of the alkali or alkaline earth metals. There are several roadcuts in the area that
contain different minerals. At this particular site, very nice augite crystals can
be found by searching the purple volcanic rubble leading up toward the interstate. Wherever you choose to rockhound, keep an eye out for rare specimens
of amethyst in small vugs of the host rock.

Sites 26–30

26. Elliston Agate

Agate nodules from Elliston

See map on page 71.

Land type: Mountains

GPS: N46° 24.93333' / W112° 29.78333'

Best season: Summer through fall

Land manager: USFS, Helena National Forest

Material: Quartz, agate, quartz crystals

Tools: Rock hammer

Vehicle: Any

Accommodations: Camping and RV parking on-site in Helena National Forest; camping, RV parking, and motels in Helena

Special attractions: None

Finding the site: From US 12 in Elliston, drive east 0.5 mile, turn right onto FR 227, and drive 13.5 miles to the Kading Campground. Park at the farthest end of the campground and follow the trail at the end of the road to the hilltops of the surrounding mountains.

Searching for agate nodules

Rockhounding

Halfway between Helena and Deer Lodge is the last place you would expect to find nice specimens of agate—and only the trained eye can spot the cobble-like agate geodes in this area. Even then, a 1.5-mile strenuous uphill hike is required to reach the locality.

The agate nodules are quite small, usually a little larger than 1 inch in diameter, and difficult to see. Specimens are found loose in the soil and can occur along the trail, but the most concentrated locations are on the very top of the mountains. Split nodules are commonly found and seem to bunch up in concentrations with solid specimens. The nodules contain agate in solid colors of white and gray and banded agate in white, gray, brown, and black. The specimens can be worth the hike though, as many of them contain geode-style quartz crystals inside banded agate similar to the Montana dryhead agate.

27. Montana City Jasper

Montana City jasper with quartz crystal vug

See map on page 71.
Land type: Mountains, roadcut
GPS: N46° 32.593' / W111° 56.895'
Best season: Any
Land manager: Jefferson County
Material: Jasper, agate, quartz crystals
Tools: None
Vehicle: Any
Accommodations: Camping, RV parking, and motels in Montana City and Helena
Special attractions: The state capitol in Helena, which was constructed in 1899, is worth a visit. It is faced with native sandstone and granite, with a dome made of Montana copper. Across the street from the capitol are the Montana Historical Society's museum, library, and archives. The museum displays some of the early mining equipment and ores found at nearby mining camps.

Finding the site: From I-15 just south of Helena, take exit 187 (Montana City) and go to the frontage road on the west side of the interstate. The roadcut is just north of the exit on the left side of the frontage road. It is best to park your car at the gas station that is located immediately across the frontage road when you exit the freeway and walk the short distance north to the first small, humble hill on the left.

Rockhounding

The plume jasper to be found in this little hill is rather impressive. Beautiful large chunks, many more than 10 inches in diameter, are visibly eroding away from the cut. The stunning jasper occurs in several shades of brown, red, and a nice caramel color. With some luck and time, you can also find interesting samples of attractive plume agate. You can also find smaller pockets of quartz crystals and drusy within the jasper.

28. Spokane Bar Sapphire Mine

See map on page 71.

Land type: Hills

GPS: N46° 40.06560' / W111° 49.80540'

Best season: Open 9 a.m. to 5 p.m. 7 days a week from late spring through early fall; winter hours vary.

Land manager: Private, Spokane Bar Sapphire Mine

Material: Sapphire, garnet, ruby, topaz, moonstone, quartz, agate, jasper, and gold

Tools: Provided

Vehicle: Any

Accommodations: Camping and RV parking nearby on Canyon Ferry Reservoir; camping, RV parking, and motels in Helena

Special attractions: The state capitol in Helena, which was constructed in 1899, is worth a visit. It is faced with native sandstone and granite, with a dome made of Montana copper. Across the street from the capitol is the Montana Historical Society's museum, library, and archives. The museum displays some of the early mining equipment and ores found at nearby mining camps.

Lifelong sapphire miner Tye Cumley at the digging area

Finding the site: From Helena take MT 280 (York Road) east toward the Missouri River/Canyon Ferry Lake and turn right onto Hart Lane at mile marker 8. After the road makes a 90-degree jump, turn left onto Castles Road. Follow the signs to the mine.

Rockhounding

Sapphires were first discovered in 1865 in terrace gravel deposits along the Missouri River northeast of Helena that were being worked for placer gold. They were a by-product that clogged the riffles in sluice boxes and dredges, but when their value was recognized, mining the sapphires became almost as important as mining the gold. The development of synthetic corundum virtually eliminated the need for natural corundum as a source of abrasives, and commercial mining today is primarily restricted to gem-quality stones.

The Spokane Bar Sapphire Mine is situated on a high gravel terrace about 5 miles downstream from Canyon Ferry Dam and west of Hauser Lake. The mine is open year-round to the public for fee-digging for sapphires. The largest sapphire on record found at the mine was 155 carats; the largest gem-quality sapphire found was 50 carats. Anything you find in your purchased gravels at the mine is yours to keep; the rock shop proudly displays an article on the wall about a customer who found a sapphire at the mine worth several thousand dollars! You can opt to dig from the open pit, buy pay gravel, or take a longer tour to the richest deposits located off-site.

Other treasures are found as you wash and pick through the gravel. Tiny samples of ruby, garnet, topaz, moonstone, citrine, quartz, agate (often moss agate), and jasper are commonly found, and many pieces are of lapidary quality. Customers are also welcome to pan the excess dirt of their purchases for the fine gold that exists in the Spokane Bar terrace gravels.

For customers who find gem-quality sapphires, contact information for companies who offer faceting and heat-treating services through the mail is available through the rock shop. The bags of potential sapphire-bearing gravel that are for purchase are also available to be shipped. For more information contact Spokane Bar Sapphire Mine, 5360 Castles Dr., Helena, MT 59602; (406) 227-8989 or (877) DIGGEMS; sapphiremine.com.

29. Montana Gold Claims

See map on page 71.

Land type: Mountains, Rivers

GPS: N46º 11.404' / W111º 38.405'

Best Season: Spring through fall

Land Manager: Gold Prospectors Association of America

Material: Gold

Tools: Gold pan, trowel, fleck flask

Vehicle: Any

Accommodations: Camping and RV parking on Canyon Ferry Lake and Helena National Forest; camping, RV parking and motels in Helena.

Special Attractions: The state capitol in Helena, which was constructed in 1899, is worth a visit. It is faced with native sandstone and granite, with a dome made of Montana copper. Across the street from the capitol are the Montana Historical Society's museum, library and archives. The museum displays some of the early mining equipment and ores found at nearby mining camps.

A gold panning sieve is useful to separate the larger rocks from panning material—never forget to check for nuggets before dumping the cobbles. Photo by the author.

Finding the Site: In order to access this location you must register with the Gold Prospectors Association of America. The GPAA offers access to their claim in this highly-rated location as well as across the state for members of the GPAA only. Contact the association for more information. Always carry your membership card and the information for the mining claim when prospecting on a GPAA claim.

Rockhounding

This site is marked at one of the most highly rated Gold Prospector Association of America's claims in Montana. If you are serious about prospecting gold you will want to become a member of the GPAA. With a membership you are allowed access to prospect GPAA claims within the state of Montana—and the entire country.

If you want to see color in your pan, this membership is worth every penny. If you want to find a nugget, upgrade to a package that includes the gold-seeking metal detector. Either way your membership (basic was $100 in 2017) includes everything you need to gold prospect including a gold pan, claim guide and access to the many stakes in Montana, including this one know as Hopeless #1, a vast 140 acre claim within the Helena National Forest open for GPAA members to gold pan.

To join the highly recommended GPAA visit www.goldprospectors.org or contact your local GPAA chapter and become a part of the gold fever community. Montana chapters include the Central Montana Prospectors Group in Great Falls, the Blackfoot River Chapter in Lincoln, the Headwaters Chapter in Belgrade or the Yellowstone Prospectors in Billings.

30. Basin Barite

See map on page 71.
Land type: Mountains, roadcut
GPS: N46º 16.021' / W112º 18.661'
Best season: Any
Land manager: Jefferson County
Material: Barite crystals
Tools: None

Linda Bruner, co-owner of A&L Shoppers Pawn and Rock Shop in Butte, displays the large clusters of golden barite crystals she found in Basin.

Vehicle: Any

Accommodations: Camping and RV parking nearby in Helena National Forest and Beaverhead-Deerlodge National Forest; camping, RV parking, and motels in Butte

Special attractions: The World Museum of Mining and Hell Roarin' Gulch just outside of Butte have displays on mining history and a reconstructed 1800s mining town with more than 50 buildings. On the grounds of Montana Tech University in Butte is the Mineral Museum, which has more than 1,500 minerals on display, including what may be the largest gold nugget ever found in Montana.

Finding the site: From I-15 west of Boulder, take Basin exit 156 to the frontage road on the south side of the freeway. Drive 3.7 miles west on the frontage road to a large towering rock outcrop on the left side of the road and park in one of the turnouts.

Rockhounding

Although there are numerous gold mines in the general vicinity of Basin, some of the more interesting collecting can be done just a few miles west of Basin at Indian Head Rock. At this site large specimens of golden barite crystals can be found. The crystals average about half an inch in length and are so abundant in loose chunks of rock eroding from the cliff that tools aren't even necessary. Gem–quality crystals aren't likely to be found, but the crystals that are here are easy to find and may be several inches long.

Sites 31 and 32

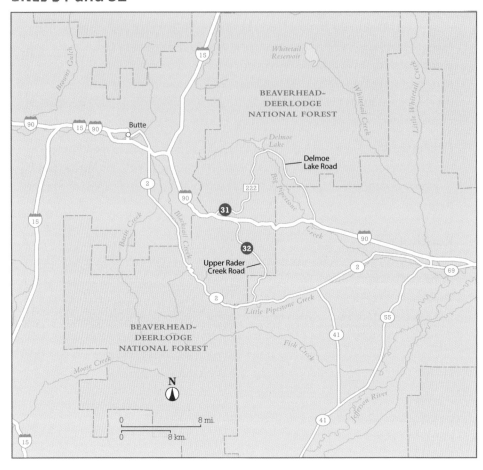

31. Homestake Pass Pegmatites

Homestake Pass pegmatites

See map on page 82.

Land type: Mountains

GPS: N45° 55.517' / W112° 23.952'

Best season: Spring through fall

Land manager: USFS, Beaverhead-Deerlodge National Forest

Material: Quartz, schorl, feldspar, garnet, mica

Tools: Rock hammer

Vehicle: Any

Accommodations: Camping and RV parking nearby in Beaverhead-Deerlodge National Forest at Delmoe Lake; camping, RV parking, and motels in Butte

Special attractions: The World Museum of Mining and Hell Roarin' Gulch just outside of Butte have displays on mining history and a reconstructed 1800s mining town with more than 50 buildings. On the grounds of Montana Tech University in Butte is the Mineral Museum, which has more than 1,500 minerals on display, including what may be the largest gold nugget ever found in Montana.

Finding the site: Pegmatites and loose large shaft crystals can be found throughout the Homestake Pass and Delmoe Lake area. To collect the best specimens, use a map as a guide and spend a day or two exploring the area. The site mentioned is the closest known locality to the highway.

From I-90 southeast of Butte, take exit 233 (Homestake). Go to the frontage road on the north side of the interstate and follow the road that leads toward Delmoe Lake (FR 222). Drive 0.8 mile and turn left onto an unmarked road. Drive 0.1 mile and park. Look for the hill on the right side of the road that shows evidence of digging—a good place to start the search.

Rockhounding

In the pegmatites of the Boulder batholith, nice specimens of quartz, schorl, feldspar, and garnet can be found. Pegmatites are an especially coarse-grained igneous rock (in this case the composition is granite) that often contain large amounts of minerals due to the rock's formation during the final and most hydrous stage of the cooling of the batholith when the magma is crystallized. The best way to search for a nice sample of Homestake pegmatites is to look for an area where others have been digging and start digging. Splitting the rock in and around the worked areas can also prove to be productive, but it all depends on the luck of the area you work. There is no shortage of virgin places to go prospecting either.

The feldspar and garnet are not defined crystals, but the chunks can be quite large. There have been reports of schorl crystals several inches long, although none have been personally observed. It seems that most rock hounds in the area are searching for the quality quartz crystals to be found. If any crystals are found by digging pegmatites, they will be difficult to remove from the rock without damaging them.

There is another way to search for complete quartz crystals, however. Several pegmatites with quartz crystal-lined vugs outcrop along the north shore of Delmoe Lake. The action of the waves upon these rocks apparently breaks the crystals loose. They can periodically be found in the sand and gravel along the shore during periods of low water. The crystals are not large, but they often possess nice terminations.

32. Rader Creek Crystals

Diligently digging in search of a crystal pocket

See map on page 82.

Land type: Mountains

GPS: N45° 53.246' / W112° 22.051'

Best season: Late spring through fall

Land manager: USFS, Beaverhead-Deerlodge National Forest

Material: Quartz, mica, feldspar

Tools: Rock hammer

Vehicle: High clearance, 4WD

Accommodations: Camping and RV parking in Beaverhead-Deerlodge National Forest; camping, RV parking, and motels in Butte

Special attractions: The World Museum of Mining and Hell Roarin' Gulch just outside of Butte have displays on mining history and a reconstructed 1800s mining town with more than 50 buildings. On the grounds of Montana Tech

University in Butte is the Mineral Museum, which has more than 1,500 minerals on display, including what may be the largest gold nugget ever found in Montana.

Finding the site: From the junction of MT 2 and MT 41 southeast of Butte, take MT 2 4.6 miles west of the junction and turn right onto Toll Mountain Road. Drive 1.3 miles and turn right onto Upper Rader Creek Road (FR 240). Continue for 4 miles to a small grove of trees on the right (1.9 miles after the intersection of Whiskey Gulch Road). On the right behind the trees, a small clearing about 200 feet into the grove exposes a well-worked quartz deposit. This is only one starting point for collecting in the Toll Mountain area; with a forest service map, you could explore the area.

Rockhounding

The rocks in the Rader Creek area between Toll Mountain and I-90 have produced some of the finest smoky quartz crystals found in the state. The quality crystals are several inches long in size and can sometimes be found in a very dark smoky shade. The crystals occur in pockets (miarolitic cavities) throughout the granite-like rocks of the Boulder batholith, which is exposed in this region. The area is a popular one for local rock hounds, and it would help to seek additional information from local rock shops or rock-club members as to its location and availability.

The easiest way to search for crystals in the Rader Creek area is to look for areas where other people have been digging. Good specimens tend to come from locations of quartz that is already exposed. Also be on the lookout for large pegmatites in the country rock and evidence that someone else has been collecting out of the same location. Oftentimes a pick and shovel are necessary to break anything free and, due to the size of most of the pits, hand tools larger than a rock hammer are recommended. The forest service does not appreciate digging, and it is possible to collect from the surface around these deposits.

Sites 33–35

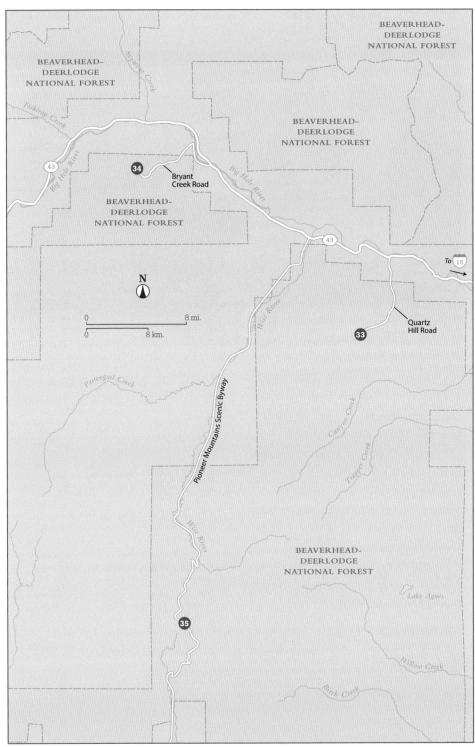

BEAVERHEAD-
DEERLODGE
NATIONAL FOREST

BEAVERHEAD-
DEERLODGE
NATIONAL FOREST

BEAVERHEAD-
DEERLODGE
NATIONAL FOREST

Seymour Creek

Fishtrap Creek

43

Big Hole River

34

Bryant
Creek Road

Big Hole River

BEAVERHEAD-
DEERLODGE
NATIONAL FOREST

43

To 15

N

Wise River

0   8 mi.

0   8 km.

Quartz
Hill Road

33

Pattengail Creek

Pioneer Mountains Scenic Byway

Canyon Creek

Trapper Creek

Wise River

BEAVERHEAD-
DEERLODGE
NATIONAL FOREST

Lake Agnes

35

Willow Creek

Birch Creek

33. Quartz Hill

Quartz Hill minerals and crystals

See map on page 87.
Land type: Mountains
GPS: (41) N45° 42.96667' / W112° 53.80000'
Best season: Summer through fall
Land manager: USFS, Beaverhead-Deerlodge National Forest
Material: Quartz, quartz crystals, malachite
Tools: Rock hammer
Vehicle: High clearance
Accommodations: Camping and RV parking in Beaverhead-Deerlodge National Forest; camping, RV parking, and motels in Butte
Special attractions: None

Finding the site: From MT 43 at the west side of Dewey, turn south onto Quartz Hill Road. Drive about 5.5 miles to the old mine tailings and an additional 1.5 miles to Quartz Hill.

Rockhounding

The Quartz Hill mining district produced significant amounts of ore from silver-bearing quartz veins spliced through the Paleozoic limestone. Along with silver, the primary mineral mined, the district also produced gold, copper, lead, and zinc. The tailings piles are the only thing left to tell the tale of this once-bustling area, and even they have little to say. Suitably, quartz and quality quartz crystals are the most abundant specimens to be collected here. The piles produce some small samples of crystals, but for the best material and the largest crystals, you'll have better luck using a forest map to explore several small forest roads that lead up to the top of Quartz Hill. Nice samples of quartz and crystals seem to be along these roads if you chase the vertical veins up the hills.

34. Calvert Loop Minerals

See map on page 87.

Land type: Mountains

GPS: N45° 50.83333' / W113° 09.18333'

Best season: Summer

Land manager: USFS, Beaverhead-Deerlodge National Forest

Material: Epidote crystals, garnet, calcite crystals, calcite, aquamarine, scheelite, pyrite

Tools: Rock hammer, sledgehammer, chisels

Vehicle: High clearance

Accommodations: Camping and RV parking in Beaverhead-Deerlodge National Forest; camping, RV parking, and motels in Anaconda and Butte

Special attractions: The World Museum of Mining and Hell Roarin' Gulch just outside of Butte have displays on mining history and a reconstructed 1800s mining town with more than 50 buildings. On the grounds of Montana Tech University in Butte is the Mineral Museum, which has more than 1,500 minerals on display, including what may be the largest gold nugget ever found in Montana.

Finding the site: From the junction of MT 43 and MT 569 between Wisdom and Wise River, drive east on MT 43 for 3.5 miles and turn right into the Dickey Bridge

Calvert loop minerals

Recreation Area. Drive past the camping and picnicking areas and stay on Bryant Creek Road/ FR 1213 for 4.2 miles to a fork in the road—this is the Calvert Loop. Stay right at the fork and drive 2.5 miles to the site. There is a large pit filled with water on the right side and a field of dump piles on the left. Park at the farthest side and explore.

Rockhounding

The Calvert tungsten mine, and the small Wise River Mining district itself, is a good example of the search for nonprecious metals after the slowdown of the Montana gold and silver rush. Mining at Calvert took place periodically on a relatively small-scale basis from 1956 to 1966. Massive dumps are spread out as testament to the 113,000 tons of ore that were mined during the ten-year run.

There is a tremendous amount of interesting rocks and minerals to be found at the Calvert mine. The mine itself looks more like an old quarry. Today a large pool of water covers the deep pit where part of a mountain once stood. Acres of old dump piles offer a great opportunity for rock hounds to collect large epidote crystals and large but fractured garnet samples. The pieces are all loose. Calcite crystals are found throughout the dumps and the area. With luck, large chunks of white calcite can be diligently split to reveal the most precious gem of them all: aquamarine, some of gem quality and up to several inches long. On one occasion the Butte Mineral and Gem Club was at the site with some nice pieces of aquamarine they discovered. Splitting the calcite is not easy, and a sledgehammer along with a nice set of chisels is needed. The prospect of finding precious gems might make the extra work worth a try.

35. Crystal Park

See map on page 87.
Land type: Mountains
GPS: N45° 29.23333' / W113° 06.03333'
Best season: Crystal Park is generally open May 15 through Oct 15.
Land manager: USFS, Beaverhead-Deerlodge National Forest
Material: Quartz crystals
Tools: Shovel, screen
Vehicle: Any
Accommodations: Camping and RV parking nearby in Beaverhead-Deerlodge National Forest; camping, RV parking, and motels in Dillon
Special attractions: Historic Bannack State Park is about 25 miles south of Crystal Park on the way to Dillon. The preserved ghost town of Bannack was Montana's first territorial capital and one of the original gold-rush towns. Today there is a visitor center and several historic buildings, as well as many tailings piles and mining equipment.

Finding the site: From just west of Wise River, drive south on the Pioneer Mountains Scenic Byway for 27 miles and turn right at the sign for Crystal Park. Park at the parking area and don't forget to pay your daily (small) per-vehicle entrance fee. If you're coming from the southern entrance, take MT 278 to the Scenic Byway, drive north 17.8 miles, and turn left at the sign for the park.

Rockhounding
Peaceful meadows, beautiful forests, and majestic peaks provide an ideal setting for Crystal Park, where rock hounds may dig for quartz crystals. The crystals were formed as a result of the Pioneer batholith, intruding up around 68 million years ago and creating the Pioneer Mountains. After a lengthy time

of cooling, the crystals formed within small pockets in the granite. Thanks to millions of years of weathering and erosion, the crystals can be found already weathered from the granite and settled loose in the soil.

The many pits occurring on the slopes suggest this locality is popular with rock hounds. The area is extensive; at the time I visited, more than 30 acres were open to the public to explore. Crystals of clear and smoky quartz and even amethyst can be found with conscientious effort. They are found only by digging in the hillside, and this means work.

The serious collector should plan to bring a shovel, pick, and some assorted mesh screens to sift the dirt and loose rock. The crystals are scattered throughout the soil in the park, and a good place to start collecting is in a pit that was already worked. Sometimes reddish layers in the soils indicate a more-condensed collection of crystals. Extremely nice crystals have been taken from this locality, some several inches long, but quality comes with time and patience. Crystal Park is open for day use, and at least one full day of exploration is recommended.

Several nearby campgrounds offer adequate facilities for those who wish to remain for a spell and enjoy some very beautiful scenery, as well as a "good dig." Crystal Park itself has picnic tables and well-maintained restroom facilities. This is a wonderful site for children; nice paths lead to the digging areas, playing in dirt is always fun, and no one seems to go home empty-handed.

It is hoped that those who visit Crystal Park will not abuse their privilege. Though the area is relatively clean, there is evidence of littering and vandalism, and some trees have fallen over due to careless digging under the roots in violation of posted rules. A concerted effort on the part of those who visit here will ensure a continuing source of fine crystal specimens for future generations. For more information contact the Beaverhead–Deerlodge National Forest or visit www.fs.usda.gov/activity/bdnf/recreation/rocks-minerals.

Sites 36 and 37

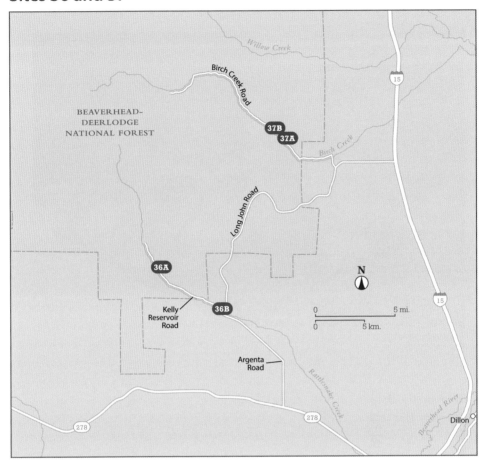

36. Argenta Minerals and Fossils

Argenta minerals

See map on page 93.

Land type: High hills

GPS: Site A: N45° 17.13333' / W 112° 52.36667'; Site B: N45° 18.71667' / W 112° 55.60000'

Best season: Spring through fall

Land manager: Bureau of Land Management

Material: Calcite crystals, pyrite, malachite, jasperoid, chert, quartz, fossils

Tools: Rock hammer

Vehicle: Any

Accommodations: Camping, RV parking in Beaverhead-Deerlodge National Forest; camping, RV parking, and motels in Dillon

Special attractions: Bannack State Historic Park is about 20 miles west of Dillon on MT 278. The preserved ghost town of Bannack was Montana's first territorial capital and one of the original gold-rush towns. Today there is a visitor center and several historic buildings, as well as many tailings piles and mining equipment.

Finding the site: From I-15 in Dillon, drive south for 2.5 miles and take exit 59 to MT 278. Drive 6.7 miles and turn right onto the road leading to Argenta. Continue 5.6 miles, turn right onto a dirt road, and drive 0.1 mile to another dirt road and turn right. If you're in a car without high-clearance four-wheel drive, park here and walk up the road 0.2 mile to the tailings piles; otherwise, drive up the road and park at one of the many mines. Be aware of open shafts at the top of the hill. To reach Site B, continue west on Thief Creek Road for 3.2 miles from the mine and turn right onto FR 606. Drive 0.1 mile to the scree slope on the left.

Rockhounding

Mining did not begin in the Argenta area until Bannack had been fully explored. Unlike the Bannack region, however, silver and lead were the more important metals produced from the Argenta District. Also, very little placer activity took place, so the vast majority of the production came from the lode deposits.

The type of rock exposed in the Argenta area is similar to that exposed at Bannack. It is sedimentary but in general much older. Most of the ore bodies occur in these rocks where mineral–bearing solutions were apparently injected either into fissures or along bedding planes, but some contact metamorphism deposits do exist within the district.

Watch out for pits that may show signs of activity, and be on the lookout for posted claims before collecting mineral specimens here. Much of the ore mined has been extremely oxidized, but certain sulfides as well as various copper carbonates occur on the dumps and tailings of the old mines. You can also find specimens of marine fossils exposed in the limestone in the region. Some of the limestone is slightly metamorphosed into marble, and the fossils are highly recrystallized. You can still identify nice examples of crinoids and bivalves.

37. Farlin Minerals

See map on page 93.
Land type: Mountains
GPS: Site A: N45° 23.65000' / W 112° 48.83333'; Site B: N45° 23.81667' / W 112° 49.25000'
Best season: Summer through fall

Farlin minerals

Land manager: USFS, Beaverhead-Deerlodge National Forest
Material: Marble, calcite, calcite crystals, glass, malachite
Tools: None
Vehicle: Any
Accommodations: Camping, RV parking in Beaverhead-Deerlodge National Forest; camping, RV parking, and motels in Dillon
Special attractions: Bannack State Historic Park is about 20 miles west of Dillon on MT 278. The preserved ghost town of Bannack was Montana's first territorial capital and one of the original gold-rush towns. Today there is a visitor center and several historic buildings, as well as many tailings piles and mining equipment.

Finding the site: From I-15 in Dillon drive north about 12 miles and take the Birch Creek exit (exit 74). Head west onto Birch Creek Road 7 miles to the old quarry on your right that leads up a hill 0.1 mile to the quarry. The second location occurs at the sign for Farlin that is reached continuing down Birch Creek Road, past the quarry, to the sign on your left.

Rockhounding

Farlin was a late 1800s silver- and copper-mining town in the Birch Creek Mining District of the beautiful Pioneer Mountains. The town once hosted hundreds of people, but a few buildings and the mine tailings are all that is left in the ghost town. There are interesting deposits to explore around the sign—the land is peppered with private-property parcels so watch for the signs. Site A is an old marble quarry that produces interesting crystals and drusy. Site B is a very unique deposit of man-made glass (much like obsidian) and is located around the Farlin sign. This material is due to the smelter that would pour the hot ash into the water, where it solidified into glass. You can also find examples of malachite and other copper minerals around the glass.

Site 38

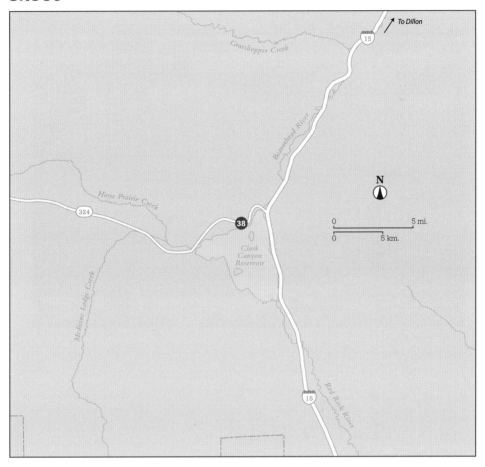

38. Clark Canyon Fossils

Clark Canyon Reservoir

See map on page 97.
Land type: Hills
GPS: (46) N44° 59.46667' / W112° 52.68333'
Best season: Spring through fall
Land manager: Bureau of Land Management
Material: Fossils, gypsum
Tools: Rock hammer
Vehicle: High clearance
Accommodations: Camping and RV parking on Clark Canyon Reservoir; camping, RV parking, and motels in Dillon
Special attractions: Bannack State Historic Park is about 20 miles west of Dillon on MT 278. The preserved ghost town of Bannack was Montana's first territorial capital and one of the original gold-rush towns. Today there is a visitor center and several historic buildings, as well as many tailings piles and mining equipment.

Finding the site: From I-15 in Dillon, drive south about 30 miles and take exit 37 to the west side of the interstate. Drive north on Old Armstead Road for 2.8 miles and stay left at the fork. At 0.3 mile after the fork, turn left onto CR 324and drive 0.7 mile to two-track dirt roads on the right. Both of these roads enter the BLM land. A high-clearance vehicle is recommended for these roads; otherwise, walk 1 mile down the BLM road to various limestone outcrops.

Rockhounding

This reservoir is located within Madison Limestone outcrops. The 350–million–year–old sedimentary rock resulted from a time when the area was covered with a shallow sea. The dark gray limestone is scattered around the lake area and this is one suggested starting point. You can easily locate specimens of horn corals, crinoids, and brachiopods in the loose pieces around the roadcut. Gypsum is also abundant in some washes and along the hills, and chert nodules can be found in the alluvium scattered along the roadsides and lakeshores.

Clark Canyon Reservoir marine fossils

Site 39

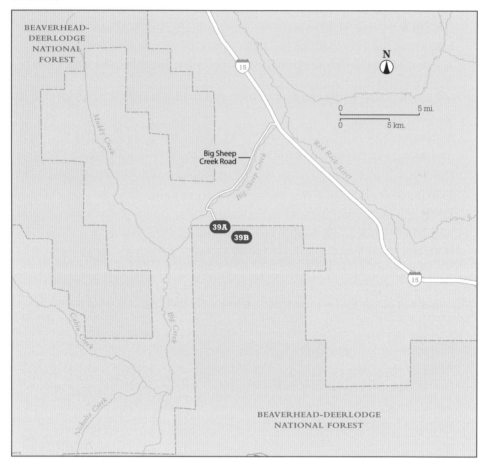

39. Dell Fossils

See map on page 100.

Land type: Mountains

GPS: Site A: N44° 39.16667' / W112° 45.18333'; Site B: N44° 38.71667' / W112° 43.91667'

Best season: Summer

Land manager: Beaverhead-Deerlodge National Forest

Material: Fossils

Tools: None

Vehicle: Any

Accommodations: Camping and RV parking on Clark Canyon Reservoir; camping, RV parking, and motels in Dillon

Special attractions: Bannack State Historic Park is about 20 miles west of Dillon on MT 278. The preserved ghost town of Bannack was Montana's first territorial capital and one of the original gold-rush towns. Today there is a visitor center and several historic buildings, as well as many tailings piles and mining equipment.

Finding the site: From I-15 take Dell exit 25 and drive to the frontage road on the west side of the interstate. Drive south on the frontage road for 1.5 miles and turn onto Big Sheep Creek Road. Drive 4.7 miles, following the signs to the Deadwood Gulch Campground. Just before the campsites, turn left onto the unlabeled road. Pass through the gate (if it is closed when you arrive, be sure to close it behind you) and park. You can begin the exploration here. The limestone occurs throughout the region, and adventurous rock hounds with high-clearance 4WD can continue farther up the road and explore the many ridges and peaks.

Rockhounding

At this location marine fossils can be found in the Beaverhead Formation's massive limestone deposits. Collecting can begin around the campground and extend across the range. The fossils are of marine critters that would have lived in the warm, shallow shelf where the limestone was formed. Keep an eye out for brachiopods, bivalves, crinoids and spectacularly large horn corals. Several specimens of loose brachiopods were found near the gate heading up the hill on the left as you drive in.

Site 40

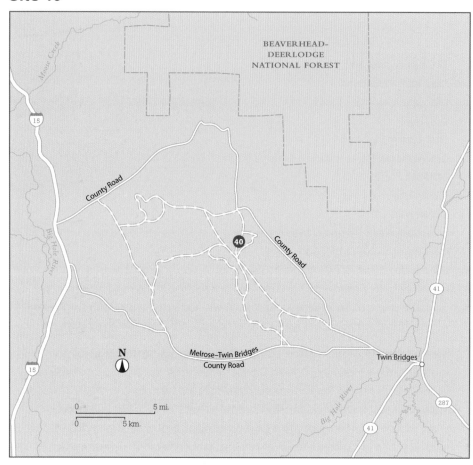

40. Rochester Mining District

See map on page 102.
Land type: Mountains
GPS: N45° 37.28333' / W112° 30.20000'
Best season: Spring through fall
Land manager: Bureau of Land Management
Material: Quartz, pyrite, mica, galena, sphalerite, malachite, azurite, chrysocolla, arsenopyrite
Tools: None
Vehicle: Any
Accommodations: Camping, RV parking in Beaverhead-Deerlodge National Forest; camping, RV parking, and motels in Dillon
Special attractions: Bannack State Historic Park is about 20 miles west of Dillon on MT 278. The preserved ghost town of Bannack was Montana's first territorial capital and one of the original gold-rush towns. Today there is a visitor center and several historic buildings, as well as many tailings piles and mining equipment.

Finding the site: From Twin Bridges drive south on MT 41 for about 1 mile, turn right onto Melrose Road, and continue as it becomes Rochester Road for about 10 miles to the Rochester Mining district. There are numerous mines and abandoned tailings piles. There is some private property within the mining area so beware of active mining claims and their postings. Most claims are posted with small metal signs on short posts near the piles. There are plenty of abandoned tailings piles to dig through without poking around the claims.

Rockhounding

The mineral deposits of the Rochester District were exploited mainly for gold and silver. The ore deposits occur primarily as veins in metamorphic rock, as contact deposits along the margins of Paleozoic limestones and igneous intrusions, or as replacement deposits in limestone. Mining of these deposits began on a small scale during the late 1800s when placer operations were attempted but proved to be too difficult to work due to the lack of water. The early exploration was quickly followed by underground mining, and during the productive years the district produced a fair quantity of high-grade ore.

The dumps and tailings, although deeply weathered, have been reported to produce, with careful searching, specimens of pyrite, galena, sphalerite, malachite, azurite, chrysocolla, arsenopyrite, quartz, and very rare specimens of cerussite, vanadinite, pyromorphite, and anglesite. At the time of my visit, nice samples of only pyrite, quartz, mica, and azurite were found, and the term "deeply" weathered can again be emphasized. If there was ever a place to view the harsh climate of Montana's rugged winters, it would be most easily described on a mineral specimen from the Rochester District.

Sites 41–46

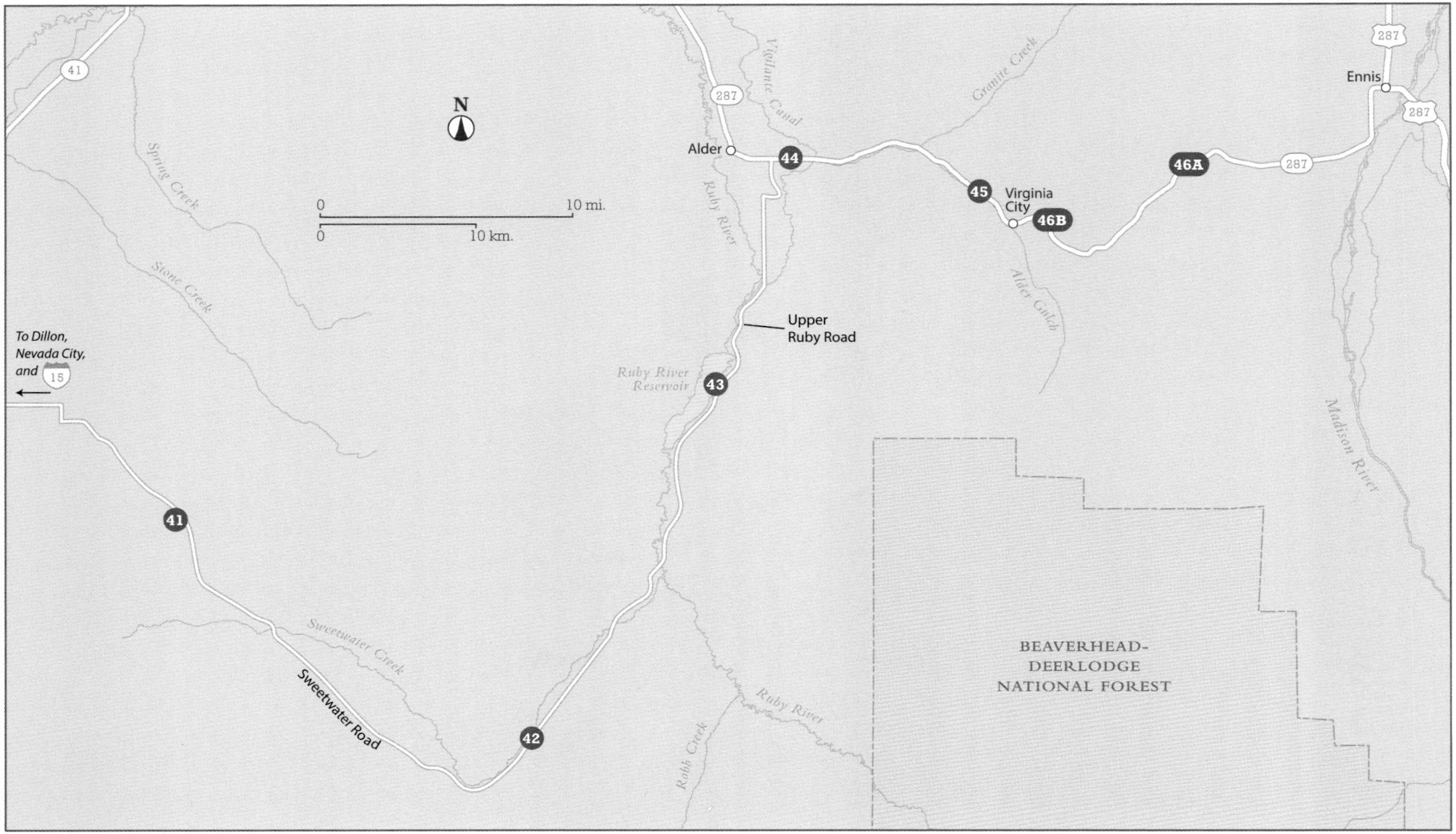

BEAVERHEAD-
DEERLODGE
NATIONAL FOREST

Madison River

Ennis

287

287

46A

46B

Virginia
City

Alder Gulch

45

Granite Creek

44

Vigilante Canal

287

Alder

Ruby River

Upper
Ruby Road

43

Ruby River
Reservoir

Ruby River

Robb Creek

42

Sweetwater Creek

Sweetwater Road

41

Spring Creek

Stone Creek

To Dillon,
Nevada City,
and   15

N

0   10 mi.

0   10 km.

41. Dillon Talc

See map on page 105.
Land type: Hills
GPS: N45° 10.21667' / W112° 25.63333'
Best season: Spring through fall
Land manager: Beaverhead County
Material: Talc
Tools: None
Vehicle: Any
Accommodations: Camping, RV parking, and motels in Dillon
Special attractions: Bannack State Historic Park is about 20 miles west of Dillon on MT 278. The preserved ghost town of Bannack was Montana's first territorial capital and one of the original gold-rush towns. Today there is a visitor center and several historic buildings, as well as many tailings piles and mining equipment.

Finding the site: From Dillon go about 1 mile north of town and turn right onto Sweetwater Road (CR 206). Drive 11.5 miles to the quarry on the right. Make sure to park away from the quarry and not to disturb quarry traffic or block the quarry's driveway. The Regal Mine is owned by Barretts Minerals Inc. and is on private land. It is uncertain if the land manager would grant permission to collect on dumps within the private property.

Rockhounding

The Regal Mine southeast of Dillon has mined talc from the open pit since 1972. The quarry is private and active. This shouldn't discourage the collector, however. There is enough interesting material to be found dumped alongside the road to keep generations of rock hounds happy. Nice specimens of soapstone, up to several pounds heavy, can be found discarded on the side of the road and in the tailings piles that line the road. The green material is quite soft and can be carved easily. When polished, it takes on the look of quality jade.

42. Sweetwater Road Rhyolite

Dylan Schmeelk searches for wonderstone.

See map on page 105.

Land type: Mountains

GPS: N45° 04.75000' / W112° 13.30000'

Best season: Spring through fall

Land manager: Bureau of Land Management

Material: Rhyolite

Tools: Rock hammer

Vehicle: Any

Accommodations: Camping and RV parking on the Ruby Reservoir; camping, RV parking, and motels in Dillon

Special attractions: Bannack State Historic Park is about 20 miles west of Dillon on MT 278. The preserved ghost town of Bannack was Montana's first territorial capital and one of the original gold-rush towns. Today there is a visitor center and several historic buildings, as well as many tailings piles and mining equipment.

Finding the site: From Alder drive south on Upper Ruby Road (MT 357) for 13.5 miles and turn right on the road leading to Dillon (Sweetwater Road). Drive 6.8

miles past the second cattle guard to large hills of eroding volcanic rock. Park on either side of the road in the small turnouts.

Rockhounding

This banded volcanic rock often referred to as "Montana Wonderstone" is a very attractive rock with bands of varying widths. Shades of yellow, brown, and red make up the colors of the bands. Technically described as silicified interbedded tuff, this material is suitable for decorative bookends and similar products. The rock does not take a high polish and therefore does not fashion well into cabochons, although some of good quality can be found. The rock was quarried as a source of terrazzo, but apparently the operation was abandoned after a short period of time. The quarry containing the rock is located on private land, and trespassing is not allowed.

A sympathetic soul or possibly a disgruntled landowner dumped several heaping piles of the material alongside the left side of the road. The reasoning behind this is unknown, but the piles are open to rock hounds. Thankfully so, and hopefully the piles will remain; otherwise, it could be difficult to obtain large pieces. The rock looks like an ordinary dark-brown to red volcanic tuff on the exposed layers; the tough tuff must be split to expose the patterned treasures. Splitting the layers requires a good deal of safety precautions. Protective eyewear, long sleeves, pants, and gloves are recommended as the chips of the rock have a tendency to pierce forcefully into the rock hound as they're split. If the piles should for some reason no longer be available along the road, small samples can be collected from splitting the tuff eroding from the roadcuts at the same location.

43. Ruby Reservoir Garnets

See map on page 105.
Land type: Lakeshore
GPS: N45° 13.56667' / W112° 07.11667'
Best seasons: Early spring or late summer through fall
Land manager: Montana Fish, Wildlife, and Parks
Material: Garnets

Jessie Dart searches for Ruby Reservoir garnets.

Tools: Screen (preferably about ⅛-inch mesh), shovel

Vehicle: Any

Accommodations: Camping and RV parking on-site; camping, RV parking, and motels in Virginia City, Ennis and Dillon

Special attractions: The reconstructed gold-rush town of Virginia City has more than 80 buildings and a rich, wild history.

Finding the site: The only specification for selecting an area to collect garnets is to be on the reservoir at a time when the water level is low. One recommendation for a nice place to park and search the shore can be reached from Alder by driving south on Upper Ruby Road (MT 357) for 7.5 miles and then turning right onto any of the several dirt roads that lead to the shore of the Ruby Reservoir.

Rockhounding

Perhaps the most impressive material to be found in the Virginia City area is garnet, which abounds in some of the dark metamorphic rock in the region. And no rock hammer will be required for the best way to search for these gems.

Gem-quality almandine garnets, suitable for faceting, are frequently found by screening the terrace and stream deposits in the valleys of the Ruby River and Sweetwater Creek. An ideal place to search for them is along the shores of the Ruby Reservoir above Ruby Dam. The garnets vary in size, although most are very small. A series of screen meshes would best separate the gravel from the fine sand, and the small gravel can be washed in the reservoir. The distinctive deep-red color of the garnets distinguishes them from the other rock particles. Garnets can also be found without screens during time of low water by walking the edges of each small terrace on the lakeshore and studying gravel deposits washed to the shore as the water level dropped. On a sunny day the garnets will shine in their deep red translucency. This method, of course, requires a bit more time than screening.

Obviously, not all the garnets found will be of gem quality, but time and practice should reward collectors with good specimens. The best time to search for garnets along the shore of the reservoir is during periods of low water. Early spring, before the streams begin carrying meltwater from the mountains, or during late summer and fall when the streams are running very low would be ideal. At no time, however, should one dig in the roadcuts along the east shore of the reservoir. The roadcuts may exhibit some rocks containing a high percentage of garnet associated with the black mineral hornblende, but the highway department has placed warning signs against rockhounding within this area. Anyone who disobeys the warning will be prosecuted.

44. Red Rock Mine Garnets

See map on page 105.
Land type: Gravel
GPS: N 45° 19.17264' / W112°3.13651'
Best season: Generally open May through Nov
Land manager: Private, Red Rock Mine
Material: Garnet, ruby, gold, feldspar, moonstone, quartz, corundum
Tools: Provided
Vehicle: Any
Accommodations: Camping and RV parking on-site; camping, RV parking, and motels in Virginia City, Ennis and Dillon
Special attractions: The reconstructed gold-rush town of Virginia City has more than 80 buildings and a rich, wild history.

Finding the site: This business is located between Alder and Virginia City on the north side of MT 287, at 2040 MT 287. They can be reached at (406) 842-5760.

Red Rock Mine gravels and polished finds

Red Rock Mine garnet

Garnet mining at Red Rock Mine

Rockhounding

This site is a personal favorite for fee mining in the state of Montana—you simply can't get more bang for your buck. Although garnets are not as prized or valuable as sapphires, this mining experience is so rich in gem-quality almandine garnets you will not be disappointed. The enthusiastic owner operates the rustic mine and has been featured on travel shows. You will likely find dozens of garnets and staff will ensure you are pleased. The mine has several gravel buckets and packages to select where you then wash and sift the gravel. All tools are provided and staff will help you organize cutting and faceting if you chose to do so with your finds. Jewelry and cut pieces from the mine are found on-site. Prices vary depending on your selections and are very low compared to sapphire mining, and even a quick stop for the lowest bucket is highly recommended.

45. River of Gold

See map on page 105.
Land type: Gravel

River of Gold

GPS: N45° 18.31667' / W111° 57.88333'

Best season: Summer

Land manager: Private, River of Gold

Material: Gold, garnet

Tools: Provided

Vehicle: Any

Accommodations: Camping and RV parking

Special attractions: Camping and RV parking on-site; camping, RV parking, and motels in Virginia City, Ennis, and Dillon

Finding the site: This site is located on the east edge of Nevada City on the south side of MT 287.

Rockhounding

This roadside gold-panning operation is a tribute to the rich history of Virginia City and also hosts mining relics. A fee for panning the pre-dug gravels goes to the historical preservation of Virginia City and Nevada City. For more information go to www.virginiacitymt.com/gold.aspx or contact River of Gold at (406) 843-5247.

46. Virginia City Minerals

See map on page 105.

Land type: Hills, roadcut

GPS: Site A: N45° 19.06667' / W111° 50.75000'; Site B: N45° 17.63333' / W111° 55.45000'

Best season: Spring through fall

Land manager: Montana DOT

Material: Quartz, garnet, feldspar, mica, zeolites

Tools: None

Vehicle: Any

Accommodations: Camping and RV parking on the Ruby Reservoir; camping, RV parking, and motels in Virginia City and Ennis

Special attractions: The reconstructed gold-rush town of Virginia City has more than 80 buildings and a rich, wild history.

Virginia City vesicular basalt zeolites

Finding the site: Several roadcuts along MT 287 between Virginia City and Ennis produce quartz, garnet, and feldspar in pegmatites and quartz vesicles in basalt. Watch for the volcanic rocks and pink, red, and black layers. Two suggested locations are the scenic overlook and roadcut at the entrance to town. Remember to park off the highway in a large turnout or on a cross road.

Rockhounding

Volcanic rocks exposed in roadcuts along MT 287 between Virginia City and Ennis host excellent examples of the difference between intrusive and extrusive igneous rocks. Intrusive igneous rocks are formed belowground as magma chambers cool. In the case of intrusive rocks, the magma has time to cool and large crystals of the various minerals can form. In this case, around the parking lot of the scenic overlook, you can find pegmatite dikes that contain white, glassy, and sometimes rose-colored quartz, as well as fine cleavable masses of pink microcline feldspar. The big pieces of feldspar have the characteristic two-directional cleavage along with chunks of quartz and large mica samples. Extrusive volcanic rocks are found at the east end of Virginia City near the sign for the entrance. Extrusive volcanic rocks are formed from magma cooling on the surface of the earth. In the case of extrusive rocks, the material cools very quickly and large minerals do not form. Here you can find vesicular basalt. The vesicles, or little holes in the rock, formed from volcanic gases as they were trapped in the rapidly cooling rock. Later, siliceous water traveled through the rock and filled the various holes with silica (quartz), which hardened into cavities lined with zeolite and carbonate minerals. Adjacent to them are dark-colored, banded metamorphic rocks containing red almandine garnet.

Site 47

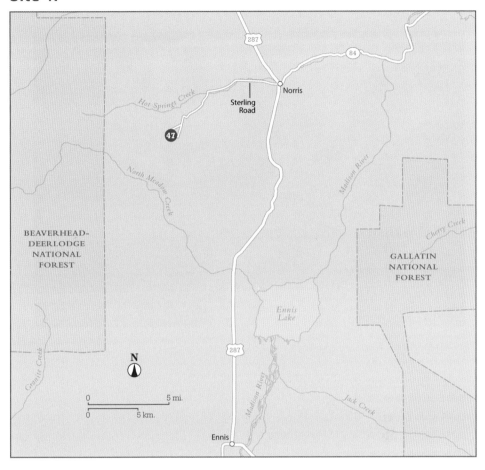

47. Norris Crystals

Norris crystals

See map on page 115.
Land type: Valley
GPS: N45° 32.35000' / W111° 47.10000'
Best season: Spring through fall
Land manager: Bureau of Land Management
Material: Quartz crystals
Tools: None
Vehicle: Any
Accommodations: Camping, RV parking, and motels in Three Forks and Ennis
Special attractions: The reconstructed gold-rush town of Virginia City has more than 80 buildings and a rich, wild history.

Finding the site: From Norris at the junction of US 287 and MT 84, turn west onto Sterling Road, the dirt road directly across from MT 84. Take the road for 3.2 miles, turn left, continue for an additional 2.7 miles, and turn right onto a small dirt road just before a cattle guard. At this point you should be able to see tailings piles; a good site is 0.2 mile down this road. Park near the pits and look around at the loose rock on the ground for quartz crystals.

Rockhounding

The quartz crystals found in the tailings piles of the mining district just west of Norris are a result of the quartz monazite intrusion that runs throughout the area. In the 1860s gold was discovered and mined on a small scale through several methods until the early 1900s. Today, small tailings piles and shallow pits spread across the barren valley are all that remain.

Nice but very small quartz crystals can be found easily in the abundant piles of white quartz tailings that cover the area. The translucent crystals are difficult to break free from the parent milky quartz, and the best samples to keep for collections are clusters of quartz crystals still attached to the rock. Splitting rocks that show potential may reveal spectacular pockets.

Sites 48–50

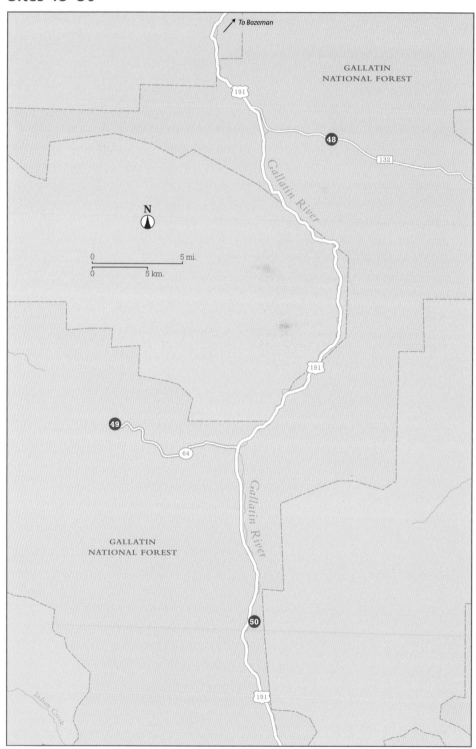

48. Spire Rock Travertine

Spire Rock travertine and crystals

See map on page 118.
Land type: Mountains, roadcut
GPS: N45° 26.205' / W111° 10.674'
Best season: Spring through fall
Land manager: USFS, Gallatin National Forest
Material: Travertine, quartz, jasper
Tools: None
Vehicle: Any
Accommodations: Camping and RV parking on-site in Gallatin National Forest; camping, RV parking, and motels in Bozeman and Big Sky
Special attractions: The Museum of the Rockies, on the grounds of Montana State University in Bozeman, hosts spectacular geology and paleontology. Yellowstone National Park is home to some of the world's premier hydrothermal and geologic features.

Finding the site: From Bozeman drive south on US 191 for 16.5 miles and take FR 132 across a small bridge to the east. Follow 132 to the right after you cross the bridge for 4.2 miles to the outcrop on the left (1.5 miles past the Spire Rock Campground). Interesting collectible rocks of different varieties line the road; this leads directly to the best travertine deposit.

Rockhounding

Rubble andesite from Eocene time covers the mountains surrounding the Squaw Creek area and much of the southern Gallatin Range in general. The same andesite volcanoes that produced the rocks containing the drusy quartz were also responsible, at some point in time, for the complete mudflow burials of the area that produced the Gallatin Petrified Forest.

The drusy quartz, referring to the pockets in the andesite completely covered in tiny quartz crystals, is abundant and covers the large volume of andesite rubble in the area. There are also intermixed deposits of precipitates including travertine with crystal pockets and vugs of drusy crystals. The crystals serve more as a thin, sparkly blanket. This is an excellent site at which to collect exquisite garden rocks and samples for bookends or other rock-saw items. While searching the hills, keep an eye out for yellow and red jasper, which can also be found in the area and creek.

49. Big Sky Fossils

See map on page 118.
Land type: Mountains, roadcut
GPS: N45º 16.746' / W111º 21.177'
Best season: Spring through fall
Land manager: Montana DOT
Material: Fossils
Tools: None
Vehicle: Any
Accommodations: Camping, RV parking, and motels in Big Sky

Big Sky fossils

Special attractions: The Museum of the Rockies, on the grounds of Montana State University in Bozeman, hosts spectacular geology and paleontology. Yellowstone National Park is home to some of the world's premier hydrothermal and geologic features.

Finding the site: From the junction of US 191 and MT 64 in Big Sky, drive 5.9 miles west on MT 64 toward the Lone Mountain Ski Resort Area to a large roadcut and park.

Rockhounding

The Cretaceous sandstone and shale in the roadcut on the way to the Lone Mountain Ski Resort can yield nice specimens of plant fossils and trace fossils. The fossils are not very detailed and are preserved as impressions or casts in the rock. They are not abundant, but loose specimens can be found at the base of the cut, and the shales and sandstones are bedded and break easily with a rock hammer.

50. Red Cliff Crystals

Red Cliff calcite crystals

See map on page 118.

Land type: Mountains, caves

GPS: N45º 10.194' / W111º 14.442'

Best season: Spring through fall

Land manager: USFS, Gallatin National Forest

Material: Calcite crystals

Tools: None

Vehicle: Any

Accommodations: Camping and RV parking on-site; camping, RV parking, and motels in West Yellowstone and Big Sky

Special attractions: The Museum of the Rockies, on the grounds of Montana State University in Bozeman, hosts spectacular geology and paleontology. Yellowstone National Park is home to some of the world's premier hydrothermal and geologic features.

Finding the site: From Big Sky drive south on US 191 for 6.5 miles and turn left at the sign to Red Cliff Campground. Drive to the south end of the campground and park at the trailhead. Hike up the trail toward the rock cliffs and caves. The first collecting location is the most easily accessible one. It's located just a couple hundred feet up the trail in the first cave. Although only a short hike, it's extremely steep. Some material can be found at the bottom of the trail near the parking area.

Rockhounding

Impressive specimens of golden calcite crystals can be collected at the Red Cliff Campground. This is a unique location since quality crystals that are several inches long can be broken free. The area has had quite a lot of collecting pressure over time, and the larger crystals won't be collected easily from the first cave. The rock containing the crystals occurs throughout the area around the steep trail, and further exploration of cavities in caves of the cliffs in the area is recommended. Either way, after visiting this locality you should be able to come home with at least a nice specimen to add to your collections. It should be noted that bats and other forest creatures seasonally call the Red Cliff caves home, and caution should be exercised to protect their delicate habitat and quiet slumber.

Sites 51 and 52

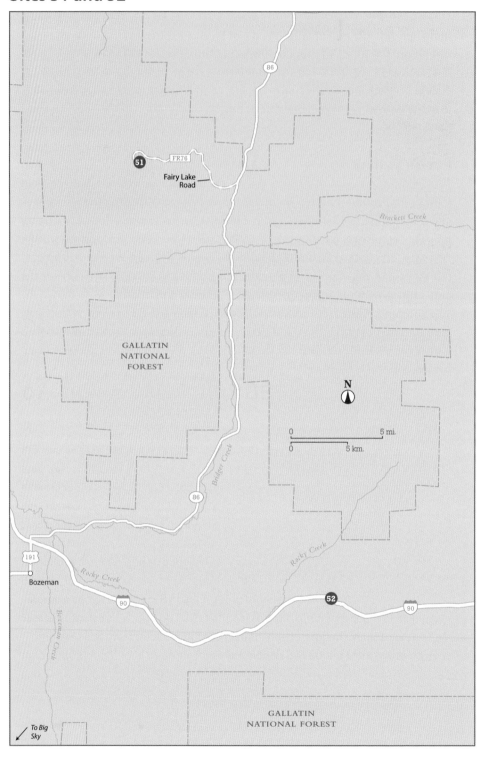

51. Fairy Lake Fossils

Beautiful Fairy Lake

See map on page 124.
Land type: Mountains, lake
GPS: N45° 54.32717' / W110° 57.45183'
Best season: Summer
Land manager: USFS, Gallatin National Forest
Material: Fossils
Tools: None
Vehicle: Any
Accommodations: Camping and RV parking on-site in Gallatin National Forest; camping, RV parking, and motels in Bozeman
Special attractions: The Museum of the Rockies, on the grounds of Montana State University in Bozeman, hosts spectacular geology and paleontology.

Finding the site: From Bozeman take MT 86 north 21 miles and turn left onto Fairy Lake Road (FR 76), a bumpy ungraded gravel road that is passable by standard vehicles but a more pleasurable experience in high clearance. Drive 6 miles to the

Marine fossils collected on the shore of Fairy Lake

trailhead and campground, where you can begin your exploration. The short, easy walk to Fairy Lake is highly recommended (not to be confused with the smaller Elf Lake directly at the parking area).

Rockhounding

This picturesque high alpine lake is lined with marine fossils. Towering gray cliffs are composed of the Mississippian Madison Limestone. The 350-million-year-old rocks were pushed up with the Bridger Range and offer a scenic backdrop true to the magical name of the lake and region. Within the limestone you can find corals, crinoids, brachiopods, bivalves, and bryozoans. The material is easily located in loose pieces along the ground and throughout the campground and hiking trails. The easy, short walk to Fairy Lake is highly recommended and specimens can be collected along the shores. A 2-mile hike up Sacajawea Peak offers additional collecting, and if you continue to the western slopes of the mountains, stromatolites can be found in even older limestones. Stromatolites are fossilized algal mats and represent the oldest continuously living thing on earth. After billions of years they remain rather unchanged and still thrive in the warm waters off the coast of western Australia.

52. Bozeman Pass Minerals

See map on page 124.

Land type: Mountains, roadcut

GPS: N45° 39.89460' / W110° 48.14580'

Best season: Any

Land manager: Gallatin County

Material: Calcite crystals, heulandite, laumontite

Tools: Rock hammer

Vehicle: Any

Accommodations: Camping and RV parking in Gallatin National Forest; camping, RV parking, and motels in Bozeman

Special attractions: The Museum of the Rockies, on the grounds of Montana State University in Bozeman, hosts spectacular geology and paleontology.

Finding the site: From Bozeman drive east on I-90 for 12 miles and take exit 319 (Jackson Creek Road) to the frontage road on the south side of the interstate. Drive 1.6 miles east to the first roadcut.

Rockhounding

Zeolite minerals are popular with many collectors and refer to any of the minerals of the zeolite group. Several zeolite minerals as well as calcite can be found within exposures of the Livingston Formation in the vicinity of Bozeman Pass.

Small calcite crystals, heulandite, and laumontite crystals are exposed in the rocks here. Unfortunately most of them occur in the area along the interstate. It is unlawful to park along any interstate except in cases of emergency, so collecting is limited to the frontage roads near the pass. With diligent searching of the road and railroad cuts east of the pass, small zeolite mineral samples can be found. The minerals are not especially abundant, but by breaking some rock, a serious collector should be rewarded with specimens that may be suitable for micromount and possibly thumbnail collections. At the particular location mentioned, nice samples of heulandite can be found.

Sites 53 and 54

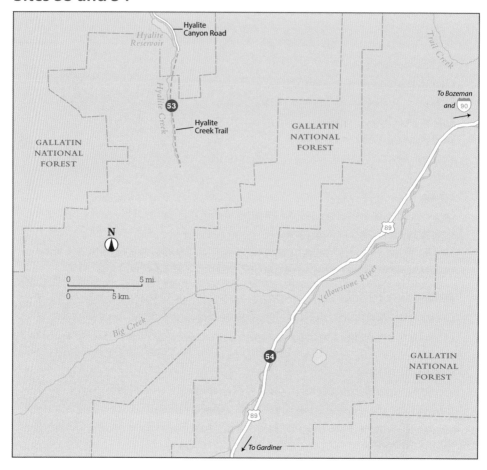

53. Hyalite Canyon

See map on page 128.
Land type: Mountains
GPS: N45° 25.75567' / W110° 57.65717'
Best season: Late spring through fall
Land manager: USFS, Gallatin National Forest
Material: Hyalite
Tools: None
Vehicle: Any
Accommodations: Camping and RV parking in Gallatin National Forest; camping, RV parking, and motels in Bozeman
Special attractions: The Museum of the Rockies, on the grounds of Montana State University in Bozeman, hosts spectacular geology and paleontology.

Finding the site: From I-90 in Bozeman, take the 19th Street exit and drive south for 9.4 miles. Turn left onto Hyalite Canyon Road and drive 13.5 miles to the Hyalite Creek trailhead. Park and pick your hiking route. This particular GPS was taken on the main trail to the falls about halfway up.

Rockhounding

This site is for the motivated rock hound, and there are several trails from which to choose. The most popular hiking trails don't offer as much material as those that gain elevation. The more strenuous trails are recommended for collecting, but the lower trails could be explored as well. Reaching the locality is a task, but for those ambitious enough to try, the scenery alone is worth the effort.

The scenic canyon derives its name from the rare presence of hyalite opal found in the rocks on Hyalite Peak at the head of the canyon. The white opal is opaque to translucent and occurs on some of the rock exposed in the area. Hyalite variety does not include the fiery, flashy opal that is well known in jewelry. The material occurs within the volcanic rocks and some can be found loose; look for the white chunks.

54. Point of Rocks Agate

Point of Rocks agate

See map on page 128.
Land type: River, gravel bars
GPS: N45° 15.826' / W110° 51.786'
Best season: Spring through fall
Land manager: Montana Fish, Wildlife and Parks
Material: Agate, jasper, quartz, petrified wood
Tools: None
Vehicle: Any
Accommodations: Camping and RV parking in Gallatin National Forest; camping, RV parking, and motels in Gardiner
Special attractions: Yellowstone National Park is home to some of the world's premier hydrothermal and geologic features.

Finding the site: From Gardiner take US 89 north for 20 miles to the Point of Rocks fishing area on the right, just north of mile marker 21.

Rockhounding

Yellowstone River rocks can be found in a variety of fishing areas and gravel bars between Gardiner and Livingston on US 89. This particular location has the best gravel bars and would take days to comb. Depending on the water levels, some wading may be required to get to the best areas. You could probably take a week to explore these gravels when the water is low and they are well exposed. Agate is easily found, including large chunks of translucent moss agates.

Agate is a variety of quartz, more specifically a form of chalcedony that is very closely related to opal. Montana agate is often collected in the gravels of the Yellowstone, but it was formed long before the time of the river. Most Montana moss agate is believed to have been created about 60 million years ago during volcanic activity in what is today the Yellowstone Park area. Agate can form as volcanic flows cool; in some way or another, water rich in silica would flow through lava and fill cavities. Perhaps the liquid silica would fill just an air pocket, or maybe a cast of a buried tree limb, and eventually harden into agate. "Montana Moss Agate" is the most sought-after variety of agate in the state and is found most abundantly in the mighty Yellowstone River. "Moss" refers to dendritic patterns that can occur in the translucent stones. Various images and beautiful landscapes can sometimes be seen in the patterning of the moss agate after it has been cut and polished.

At any time of low water, agate can be commonly found throughout the gravels of the Yellowstone River. An eye must be developed to find a piece, as they often look like plain river rock due to their deceiving dull crusts. Hints of translucency can be seen in the sunlight—an instrumental tool to the search. A trip to Montana isn't complete without finding a beautiful moss agate. From pebble-size nodules to crystal-filled geodes and limb casts of trees, there is plenty to be found—perhaps the stunning piece for your next bola tie. At this site, when the water is low, you can search for these unique state gems. Jasper, petrified wood, and other varieties of quartz collectibles such as banded chalcedony also dot the Yellowstone River.

Sites 55 and 56

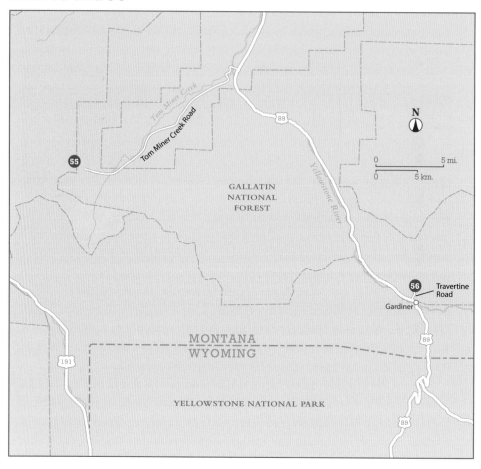

55. Gallatin Petrified Forest

Fossil tree in the Gallatin Petrified Forest

See map on page 132.

Land type: Mountains

GPS: N45° 08.105' / W111° 04.275'

Best season: Summer

Land manager: USFS, Gallatin National Forest

Material: Petrified wood, quartz crystals, citrine, amethyst

Tools: Rock hammer

Vehicle: Any

Accommodations: Camping and RV parking on-site in Gallatin National Forest; camping, RV parking, and motels in Gardiner and Livingston

Special attractions: Yellowstone National Park is home to some of the world's premier hydrothermal and geologic features.

Finding the site: From Gardiner drive north on US 89 for 16 miles to just past the town of Miner. Turn left onto Tom Miner Creek Road, stay left at the T intersection, and continue about 10 miles to the Tom Miner Campground. The trailhead is located at the bend in the campground, near the head of a beautiful mountain valley. Petrified wood can be found in the gullies and drainage of the area after hiking 1.5 miles, but the spectacular petrified tree stumps and "forests" are found several miles farther up the trail. A permit is required to collect petrified wood in the Gallatin Petrified Forest. The free permit can be obtained from the Forest Service office in Gardiner by visiting the office during their hours of operation. It is highly suggested that you visit the office even if you already have your permit. The staff offers directions, a map, and recommendations on hiking and collecting areas.

For more information or to get a permit, contact the Gardiner Ranger District at (406) 848-7375 or visit www.fs.usda.gov/main/gallatin/passes-permits/forestproducts.

Rockhounding

The fossil forests of Yellowstone National Park are unique in that they lie one atop another. Entire forests were buried by volcanic ash and stream-carried volcanic sediment. New forests would grow, only to be buried by the same processes. In time, the woody fibers of the trees in these buried forests were replaced by silica, thereby preserving the delicate cellular structures of the wood. A northern extension of these rock forests of Yellowstone can be found in southern Park County and southeastern Gallatin County. This is the area known as the Gallatin Petrified Forest.

Wind, water, and ice have carved into the thick layers of volcanic rock in which the many petrified trees have been preserved. The logs and stumps are now exposed in the cliffs and steep slopes high on the valley sides. Petrified wood no longer can be found near the campground. It was removed long ago by eager collectors. Climbing to the higher elevations and carefully searching the rocky slopes north of the campground near Ramshorn Peak will reveal pieces that can be collected, as well as some very large petrified logs and stumps that can be photographed. It is necessary to hike a fair distance over rough and relatively high-elevation terrain (more than 5,000 feet) to reach good collecting and viewing areas, so collectors should be in good physical condition.

Much of the fossil wood's state of preservation is extremely good. Frequently, the casts of limbs and logs prove to be hollow and lined with beautiful quartz crystals and, on occasion, crystals of amethyst. Some rock hounds have reported the detailed impression of leaves in the finer-grained siltstones sometimes associated with the coarser-grained rock preserving the fossil wood.

Attempts have been made in the past by greedy collectors to remove large stumps and logs. When these attempts failed, the majestic monoliths were destroyed with hammer and chisel. It is hoped that the established collecting limitations will discourage this kind of ignorance.

The federal regulations concerning the amount of petrified wood that can be collected are different from those on other public land and are strictly enforced. Only a relatively small piece of wood, 20 cubic inches, can be collected by each individual per year within the Gallatin National Forest. Exercising extreme selectivity in your collecting is encouraged. The size of the "souvenir" specimen should in no way deter one from visiting the Petrified Forest, however. The number of standing stumps and fallen logs of petrified wood are worth seeing. A camera can record for you the fantastic part of the earth's geologic history that is preserved in the rocks at this unique area.

56. Gardiner Travertine

See map on page 132.
Land type: Mountains, roadcut
GPS: N45º 02.696' / W110º 42.389'
Best season: Spring through fall
Land manager: USFS, Gallatin National Forest
Material: Travertine
Tools: None
Vehicle: Any
Accommodations: Camping, RV parking, and motels in Gardiner and Yellowstone National Park
Special attractions: Yellowstone National Park is home to some of the world's premier hydrothermal and geologic features.

Finding the site: From Gardiner drive northeast on 4th Street and make a quick right onto Jardine. Drive 1 mile and turn left at the fork in the road onto Travertine Road (not to be confused with Travertine Street on the north end of town), drive 1.3 miles, and turn left and explore the roadcuts. Do not trespass into the active mining areas; there is plenty of material to be found along the road.

Gardiner travertine

Rockhounding

Although Gardiner is most noted for being the north entrance to Yellowstone National Park, the vicinity has some interesting geological features. One of these is an extensive deposit of travertine, closely related to that being formed by hydrothermal activity at Mammoth Hot Springs south of Gardiner, just within the boundary of the park. Travertine is the name given to the layered, vuggy limestone that owes its origin to hot-spring deposition. It forms when the hot water carrying dissolved calcium carbonate reaches the surface, cools, and evaporates, thereby causing the mineral material to precipitate. Over long periods of time, thick deposits of travertine can accumulate.

Several sites are presently being quarried, and the material is shipped out to factories, where it is shaped into attractive blocks and pieces for buildings, patios, and the like. Access to active quarries is of course questionable, but roadcuts high in the hills surrounding the quarries can provide rock-garden-variety pieces with less color. The collectible pieces tend to be white, red, orange, and pink banded, some with nice pockets or "vugs" of small calcite crystals. The travertine cuts nicely with a rock saw, and attractive bookends with vugs of crystals can be made.

Site 57

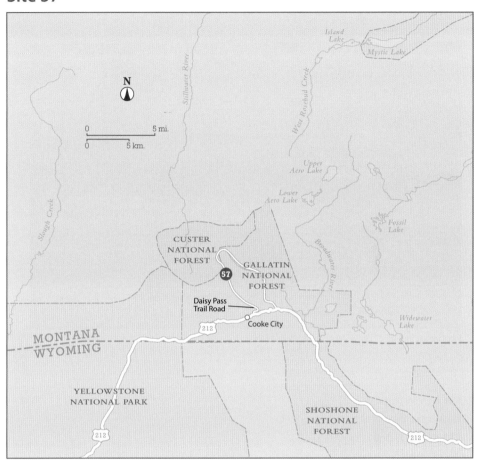

57. Daisy Pass Minerals

Sweeping scenery of Daisy Pass

See map on page 137.

Land type: Mountains

GPS: N45° 03.163' / W109° 57.072'

Best season: Summer

Land manager: USFS, Gallatin National Forest

Material: Pyrite, magnetite, chalcopyrite, garnet, quartz crystals

Tools: None

Vehicle: High clearance, 4WD

Accommodations: Camping, RV parking, and motels in Cooke City and Yellowstone National Park

Special attractions: Yellowstone National Park is home to some of the world's premier hydrothermal and geologic features. Beartooth Pass, which crests at more than 10,000 feet, an elevation well above the tree line, and Grasshopper Glacier, where you can view locusts that were trapped in the glacial mass, are also nearby.

Finding the site: From Cooke City the New World Mining district can be reached via several rugged mountain roads. The 5 miles from Cooke City up to Daisy Pass are barely negotiable by regular automobile. A high-clearance four-wheel-drive vehicle is strongly recommended for the roads in this area. Some of the most scenic country of the nation is situated here. The trails leading from Cooke City to the old mining areas offer access to those who wish to explore the Beartooth Mountains further by backpack. The easiest directions to the site are to drive to the east end of Cooke City and turn left, heading north, onto the road connecting to Daisy Pass Trail Road. Go about 5 miles to the first mountainside covered in mine tailings.

Another highly recommended and popular option for those without a proper vehicle is to rent an ATV in Cooke City.

Rockhounding

High in the mountain passes north of Cooke City exists an area in which silver and lead were discovered in relatively large amounts during the early 1900s. In the vicinity of Daisy Pass, numerous mine dumps and tailings tell of a period when mining activity was much more pronounced. Exposed here are Precambrian metamorphic rocks overlain by lower Paleozoic sedimentary rocks and Tertiary volcanic rocks. Tertiary intrusions in the form of laccoliths, sills, and dikes occur, and the mineralization in the area is thought to be related to these. Dumps and tailings support considerable amounts of pyrite, although most of it is considerably weathered. Magnetite is found along with some copper minerals. On a sunny day the rocks sparkle with the various metallic minerals and crystals streaming down the mountains.

Sites 58–61

58. Red Lodge Fossils

See map on page 140.

Land type: Hills

GPS: N45º 09.478' / W109º 10.562'

Best season: Spring through fall

Land manager: Montana DOT

Material: Fossils

Tools: None

Vehicle: Any

Accommodations: Camping, RV parking, and motels in Red Lodge

Special attractions: Yellowstone National Park is home to some of the world's premier hydrothermal and geologic features. Beartooth Pass, which crests at more than 10,000 feet, an elevation well above the tree line, and Grasshopper Glacier, where you can view locusts that were trapped in the glacial mass, are also nearby.

Red Lodge fossils

Finding the site: From Red Lodge take MT 308 east for 5.7 miles to a roadcut on the right. Plant fossils were collected about 0.6 mile from the memorial on the south side of the road on a roadcut along the highway at the road that leads to a tower. The tower road is private, but that is not a problem as the samples are located in the roadcuts by the highway. Park on the tower road and explore.

Rockhounding

This bittersweet site offers nice samples of fossil plants, with details of wood and leaves. The plant fossils are associated with the coal seam. In 1943 an explosion occurred at the mine due to the buildup of methane gas deep underground, and seventy-five men lost their lives. A roadside memorial pays homage to the miners and offers an overlook of the remaining operation.

59. Belfry Fossils

Searching for Belfry fossils

See map on page 140.
Land type: Hills
GPS: N45° 04.400' / W109° 01.915'
Best season: Spring through fall
Land manager: Bureau of Land Management
Material: Fossils, petrified wood, agate, jasper
Tools: None
Vehicle: Any
Accommodations: Camping, RV parking, and motels in Red Lodge and Laurel
Special attractions: None

Finding the site: From the intersection of MT 72 and MT 308 in Belfry, drive south on MT 72 for 5.6 miles. Turn left immediately after crossing the Clarks Fork of the Yellowstone River onto a dirt road leading to a river access area. Drive 0.2 mile and explore the ridges along the BLM land for fossils. If the river is low, you can also explore the gravels for agate, jasper, and petrified wood.

Rockhounding

Abundant fossils of freshwater clams can be found in the Fort Union Formation south of Belfry just before the Wyoming state line. Most of the fossils are highly weathered, but well-preserved complete clams with both sides of the shell can be found loose from the parent rock. The large specimens are found mostly in the thick slabs of sandstone that is almost impossible to break. With diligent searching, quality specimens can be found loose or in smaller chunks of rock. The concentration of exposed sandstone that contains the fossils also offers burrow casts and other trace fossils. The Clarks Fork of the Yellowstone River can additionally be explored for agate, jasper, and petrified wood.

60. Pryor Mountains Chert and Fossils

See map on page 140.

Land type: Hills, badlands

GPS: Site A: N45º 04.050' / W108º 33.470'; Site B: N45º 03.853' / W108º 32.420'; Site C: N45º 02.669' / W108º 25.031'

Best season: Spring through fall

Land manager: Bureau of Land Management

Material: Fossils, agate, jasper, chert, pyrite, calcite crystals

Tools: None

Vehicle: Any

Accommodations: Camping on-site; RV parking, and motels in Lovell, Wyoming

Special attractions: Chief Plenty Coups National Historic Monument at the Crow Indian Reservation includes a museum of Native American history. South of the rockhounding site, the Pryor Mountain Wild Horse Range offers an opportunity to view wild horses and the spectacular Big Horn River.

Finding the site: From US 310 in Warren, turn left onto Helt Road leading toward the limestone quarry. Drive 2.7 miles and turn right onto Helt Road. To reach Site A, drive 3.4 miles down Helt Road to a faint road leading to the base of a large red butte across from Bear Canyon Road. Park at the bottom and explore the butte for fossilized coral.

Site B can be reached by continuing east on Helt Road for 1 mile to a fork in the road. Park here and explore the roadside and surrounding hills for loose fossils, agate, chert, and jasper. Site C can be reached by driving 4 miles past Road 1022

Paleontologist Amy Singer and the Pryor Mountains

farther down Helt Road and turning left onto Crooked Creek Road. A nice area to begin collecting star crinoids and oysters can be found 2.7 miles down Crooked Creek Road. Search the surrounding hills and wash beds for loose fossils. Loose shale at the tops of the hills can be split easily, exposing the occasional insect.

Rockhounding

The Pryor Mountains consist of several fault blocks that were uplifted by diastrophic processes during the Late Cretaceous and Early Tertiary periods. As a result of this mountain building and subsequent erosion, sedimentary rock layers ranging in age from Early Paleozoic to Late Cretaceous occur within and on the margins of the range. Almost all of the rock layers are fossiliferous to varying degrees, and some contain unusual mineral deposits and rocks with lapidary potential.

The Madison limestone, a formation of late Paleozoic age, reveals spirifer brachiopods, coral, and crinoid remains massed together in some outcrops. Collapsed caverns within the limestone have been the source of some secondary uranium minerals. The low-grade ore was mined on a small scale in the past, and the mines have recently been reclaimed, although some uranium, calcite crystals, fluorite crystals, and pyrite can still be found where the mines once were. The uranium mineral tyuyamunite occurs as a yellow powdery coating on the calcite crystals or as disseminated grains throughout the brecciate limestone. Exposures of Late Paleozoic sediments on the flanks of the Pryors contain an abundance of limonite after pyrite or goethite after pyrite pseudomorphs. These occur as clusters of cubic crystals. Most are deeply

(From top left to right to bottom left to right) Pentacrinus columnals, the oyster Gryphaea, the coral Astrocoenia, the bullet-shaped belemnite Pachyteuthis, and samples of banded chert and jasper from exposures of marine Jurassic rocks in the Pryor Mountain region

oxidized, but some retain a very dark color with submetallic luster. Many of them, if broken, reveal original pyrite in the center.

Along the western and southern margins of the Pryors are excellent exposures of Mesozoic rocks. The bright-red Triassic Chugwater Formation is quite noticeable in the landscape but is nonfossiliferous. Marine Jurassic formations directly overlaying the Chugwater Formation, however, provide an abundance of fossils. At this site the colonial coral Astrocoenia is found. The belemnite Pachyteuthis, oysters of the genera Gryphaea and Ostrea, and the star-shaped columnals of the crinoid Pentacrinus are especially prolific.

The Pryor Mountains are a personal favorite because there are so many treasures to be found and such interesting stratigraphy to keep the geologic mind intrigued. Throughout the area is an abundance of fossils, jasper, and agate that can be found easily by exploring the ground, particularly the wash beds of the area, along with the intact layers of the hills. Your best bet is to camp on-site at the BLM Bear Gulch Campground and spend a couple of days wandering this beautiful high desert.

61. Red Dome Fossils

Emilia Palenius searches for fossils at Red Dome.

See map on page 140.
Land type: Hills, roadcut
GPS: N45º 12.423' / W108º 47.884'
Best season: Spring through fall
Land manager: Bureau of Land Management
Material: Fossils
Tools: Rock hammer
Vehicle: Any
Accommodations: Camping, RV parking, and motels in Red Lodge
Special attractions: Chief Plenty Coups National Historic Monument at the Crow Indian Reservation includes a museum of Native American history. South of the site the Pryor Mountain Wild Horse Range offers an opportunity to view wild horses.

Finding the site: From Bridger drive south on US 310 for 2.6 miles, turn left onto Pryor Mountain Road, and continue 8 miles and explore the roadcuts on the right.

A BLM map is recommended to ensure no trespassing, because the road traverses through private and public land. The area around Red Dome produces a variety of fossils; the site listed is a recommended starting point.

Rockhounding

Several fossiliferous roadcuts outcrop on the road leading to the Pryor Mountain Ice Cave. The icy cave is a hangover from the Pleistocene and, true to its name, holds ice year-round. Red Dome itself is an interesting geologic formation, and the area has been the subject for unique finds, including several species of dinosaurs. The dome is an arch of Triassic mudstones, an anticline with a striking red color that dominates the area visually. There are several localities both on private land and BLM land east of Bridger that show excellent exposures in the marine Jurassic Piper, Rierdon, and Swift Formations. These rock units contain a relative abundance of fossils, including amazing oyster conglomerates, belemnites, star crinoids columnals, and associated fauna, including several species of clams, snails, and ammonites. At the particular location listed, Red Dome rests as a phenomenal backdrop to the roadcut with marine fossils. Look for chunks of limestone oyster conglomerates loose on the hill on the right side of the road. The GPS mentioned should simply be a starting place for roadcuts.

Red Dome Gryphaea oyster fossil

Sites 62–67

62. Itch-Kep-Pe Agate

See map on page 148.

Land type: River, gravel bars

GPS: N45° 37.348' / W109° 14.179'

Best season: Summer

Land manager: City of Columbus

Material: Agate, jasper, petrified wood

Tools: None

Vehicle: Any

Accommodations: Camping and RV parking on-site; camping, RV parking, and motels in Columbus

Special attractions: Pictograph Cave National Historic Landmark south of Billings. "Pompey's Pillar" is a popular tourist attraction; the outcrop of Cretaceous sandstone along the Yellowstone River was signed by Captain Clark of the Lewis and Clark expedition in 1806.

Finding the site: From I-90 in Columbus take exit 78 south for 1.4 miles to the city park to your left (east). Park and explore the gravels.

Rockhounding

Agate is a variety of quartz, more specifically a form of chalcedony that is very closely related to opal. Montana agate is often collected in the gravels of the Yellowstone, but it was formed long before the time of the river. Most Montana moss agate is believed to have been created about 60 million years ago during volcanic activity in what is today the Yellowstone Park area. Agate can form as volcanic flows cool; in some way or another, water rich in silica would flow through lava and fill cavities. Perhaps the liquid silica would fill just an air pocket, or maybe a cast of a buried tree limb, and eventually harden into agate. "Montana moss agate" is the most sought-after variety of agate in the state and is found most abundantly in the mighty Yellowstone River. "Moss" refers to dendritic patterns that can occur in the translucent stones. Various images and beautiful landscapes can sometimes be seen in the patterning of the moss agate after it has been cut and polished.

At any time of low water, agate can be commonly found throughout the gravels of the Yellowstone River. An eye must be developed to find a piece, as they often look like plain river rock due to their deceiving dull crusts. Hints of translucency can be seen in the sunlight—an instrumental tool to the search. A trip to Montana isn't complete without finding a beautiful moss agate. From pebble-size nodules to crystal-filled geodes and limb casts of trees, there is plenty to be found—perhaps the stunning piece for your next bola tie. At this site, when the water is low, you can search for these unique state gems. Jasper, petrified wood, and other varieties of quartz collectibles such as banded chalcedony also dot the Yellowstone.

63. Laurel Agate

See map on page 148.
Land type: River, gravel bars
GPS: N45º 39.233' / W108º 45.448'
Best season: Summer
Land manager: City of Laurel
Material: Agate, jasper, petrified wood
Tools: None
Vehicle: Any
Accommodations: Camping and RV parking on-site; camping, RV parking, and motels in Laurel
Special attractions: Pictograph Cave National Historic Landmark south of Billings. "Pompey's Pillar" is a popular tourist attraction; the outcrop of Cretaceous sandstone along the Yellowstone River was signed by Captain Clark of the Lewis and Clark expedition in 1806.

Finding the site: From the town of Laurel, west of Billings, head south, turn left into Riverside Park, and explore the gravel bars.

Rockhounding

This shady riverfront park is another good location to look for one of the state gems. Agate is a variety of quartz, more specifically a form of chalcedony that is very closely related to opal. Montana agate is often collected in the gravels of the Yellowstone, but it was formed long before the time of the river. Most Montana moss agate is believed to have been created about 60 million years

Unpolished Laurel agate, chalcedony, and petrified wood

ago during volcanic activity in what is today the Yellowstone Park area. Agate can form as volcanic flows cool; in some way or another, water rich in silica would flow through lava and fill cavities. Perhaps the liquid silica would fill just an air pocket, or maybe a cast of a buried tree limb, and eventually harden into agate. "Montana moss agate" is the most sought-after variety of agate in the state and is found most abundantly in the mighty Yellowstone River. "Moss" refers to dendritic patterns that can occur in the translucent stones. Various images and beautiful landscapes can sometimes be seen in the patterning of the moss agate after it has been cut and polished.

At any time of low water, agate can be commonly found throughout the gravels of the Yellowstone River. An eye must be developed to find a piece, as they often look like plain river rock due to their deceiving dull crusts. Hints of translucency can be seen in the sunlight—an instrumental tool to the search. A trip to Montana isn't complete without finding a beautiful moss agate. From pebble-size nodules to crystal-filled geodes and limb casts of trees, there is plenty to be found—perhaps the stunning piece for your next bola tie. At this site, when the water is low, you can search for these unique state gems. Jasper, petrified wood, and other varieties of quartz collectibles such as banded chalcedony also dot the Yellowstone.

64. Dino Lab Digs

Working on a dinosaur dig in the Little Snowy Mountains

See map on page 148.
Land type: Mountains, badlands
GPS: N45° 47.35333' / W108° 29.41833'
Best season: Summer
Land manager: Private, Judith River Dinosaur Institute
Material: Fossils
Tools: Provided
Vehicle: Any
Accommodations: Provided
Special attractions: Pictograph Cave National Historic Landmark south of Billings. "Pompey's Pillar" is a popular tourist attraction; the outcrop of Cretaceous sandstone along the Yellowstone River was signed by Captain Clark of the Lewis and Clark expedition in 1806.

Finding the site: All digs begin at the Dino Lab, where the lab classes are held, in Billings at 1518 1st Ave. North. From I-90 in Billings take Exit 446 north for 5 miles and turn left onto 18th Street then right onto 1st Avenue where Dino Lab is located on the right.

Rockhounding

The Dino Lab in Billings offers several different options for those who want to have hands-on experience working with dinosaurs. You can enroll in lab preparation classes and tours around the Dino Lab or join one of their week-long digs in the beautiful Little Snow Mountains near Grass Range, Montana. The popular dinosaur digs ran $1,700/week for the 2015 season and tend to fill up. The minimum age is 12. For more information visit www.montana dinosaurdigs.com or the Dino Lab at 1518 1st Ave. North, Billings, MT 59101 (406-696-5842).

65. Riverfront Park Agate

See map on page 148.
Land type: River, gravel bars
GPS: N45° 44.325' / W108° 32.020'
Best season: Spring through summer
Land manager: City of Billings
Material: Agate, petrified wood, chert, jasper
Tools: None
Vehicle: Any
Accommodations: Camping, RV parking, and motels in Billings
Special attractions: Pictograph Cave National Historic Landmark south of Billings. "Pompey's Pillar" is a popular tourist attraction; the outcrop of Cretaceous sandstone along the Yellowstone River was signed by Captain Clark of the Lewis and Clark expedition in 1806.

Finding the site: From I-90 in Billings, exit at South Billings Boulevard. Drive south for 0.7 mile and turn left into Riverfront Park. Stay right on the road inside the park, which dead-ends at several parking spaces along the Yellowstone River.

Rockhounding

The gravels in and adjacent to the Yellowstone River south of Billings are a source of petrified wood and occasional agates. Banded chert and red jasper also frequent the gravels. The assorted siliceous rocks cut and polish nicely. The translucent stones can be found in several colors and patterns by simply walking the shore. Agate is a variety of quartz, more specifically a form of chalcedony that is very closely related to opal.

Montana agate is often collected in the gravels of the Yellowstone, but it was formed long before the time of the river. Most Montana moss agate is believed to have been created about 60 million years ago during volcanic activity in what is today the Yellowstone Park area. Agate can form as volcanic flows cool; in some way or another, water rich in silica would flow through lava and fill cavities. Perhaps the liquid silica would fill just an air pocket, or maybe a cast of a buried tree limb, and eventually harden into agate. "Montana Moss Agate" is the most sought-after variety of agate in the state and is found most abundantly in the mighty Yellowstone River. "Moss" refers to dendritic patterns that can occur in the translucent stones. Various images and beautiful landscapes can sometimes be seen in the patterning of the moss agate after it has been cut and polished.

At any time of low water, agate can be commonly found throughout the gravels of the Yellowstone River. An eye must be developed to find a piece, as they often look like plain river rock due to their deceiving dull crusts. Hints of translucency can be seen in the sunlight—an instrumental tool to the search. A trip to Montana isn't complete without finding a beautiful moss agate. From pebble-size nodules to crystal-filled geodes and limb casts of trees, there is plenty to be found—perhaps the stunning piece for your next bola tie. At this site, when the water is low, you can search for these unique state gems. Jasper, petrified wood, and other varieties of quartz collectibles such as banded chalcedony also dot the Yellowstone.

Periods of low water are always the best time to search in river gravels for potential cutting material. A bike trail at the parking area follows the river. This is also an excellent area for picnicking since it's within a large green park with facilities and a pond.

66. Billings Fossils

Billings fossils and crystals

See map on page 148.
Land type: Hills
GPS: N45° 44.140' / W108° 31.505'
Best season: Spring through fall
Land manager: Bureau of Land Management
Material: Calcite crystals, barite crystals, celestite crystals, fossils
Tools: None
Vehicle: Any
Accommodations: Camping, RV parking, and motels in Billings
Special attractions: Pictograph Cave National Historic Landmark south of Billings. "Pompey's Pillar" is a popular tourist attraction; the outcrop of Cretaceous sandstone along the Yellowstone River was signed by Captain Clark of the Lewis and Clark expedition in 1806.

Finding the site: From I-90 in Billings, exit at South Billings Boulevard and drive south for 1.2 miles. Turn left onto Old Blue Creek Road and drive 0.4 mile to the South Hills OHV Park and park.

Rockhounding

The Cretaceous Colorado Shale of marine origin makes up the high hills along the Yellowstone River in this area directly south of Billings. Throughout the formation numerous limestone concretions can be found, most of which are septarian in nature, with veins of white and brown calcite. Periodically, crystals of barite and celestite are found associated with calcite crystals in small cavities within the veins themselves. Fossils, too, are sometimes found in some of the concretions, and although pelecypods predominate, some relatively nice specimens of ammonites have been found, although trace fossils of burrows and crystals of gypsum and calcite are more commonly found.

Search the area in and around the base of the hills. Crystals can be found by either splitting chunks of limestone or searching the loose chunks at the base of the hills. Fossils are not very common. It should be noted that although this park is open for collecting crystals and fossils, it is designated as an off-road–vehicle park for motorcycles and other ATVs. The rock hound should stay in open areas and be cautious of the off-road traffic; riders risk not being able to see or hear pedestrians.

67. Stratford Hill Fossils

See map on page 148.
Land type: Hills, roadcut
GPS: N45º 35.042' / W108º 31.505'
Best season: Any
Land manager: Yellowstone County
Material: Fossils
Tools: Rock hammer, chisel, goggles
Vehicle: Any
Accommodations: Camping, RV parking, and motels in Billings
Special attractions: Pictograph Cave National Historic Landmark south of Billings. "Pompey's Pillar" is a popular tourist attraction; the outcrop of Cretaceous sandstone along the Yellowstone River was signed by Captain Clark of the Lewis and Clark expedition in 1806.

Finding the site: From I-90 in Billings, exit at South Billings Boulevard. Drive 1.7 miles south and turn right onto Hill Crest Road. Continue 4.3 miles on Hillcrest Road until it becomes gravel and forks. Stay left, drive 5.7 miles, and turn left onto Stratford Hill Road. Continue for 2 miles to a large roadcut in the highest hill on the left side of the road.

Searching for fossils at Stratford Hill.

Rockhounding

The Mowry Shale at this site contains flattened specimens and imprints of ammonites and extremely well-preserved fish scales, some of which retain their iridescence and colors. Located in the rolling hills just outside of Billings, the 20-foot-tall layers of the roadcut can be tempting to explore all day, especially since the rock is easily split and quite productive. Fossils also can be found by looking through the loose chunks at the bottom of the cut. The best specimens have been found by carefully splitting the larger pieces of the shale with a rock hammer and a good pair of goggles. The fish scales are abundant; the ammonites and imprints, especially a whole one, require a more diligent search.

Ammonites from Stratford Hill Road

Site 68

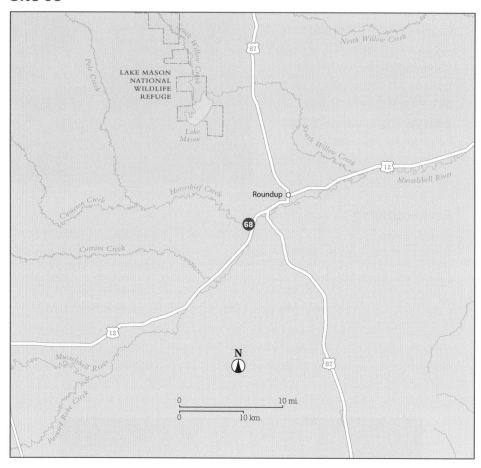

68. Klein Fossils

See map on page 158.

Land type: Low mountains, roadcut

GPS: N46° 25.21000' / W108° 35.81167'

Best season: Spring through fall

Land manager: Montana DOT

Material: Fossils

Tools: None

Vehicle: Any

Accommodations: Camping, RV parking, and motels in Roundup

Special attractions: Musselshell Valley Historical Museum in Roundup has a small selection of fossils as well as historical relics.

Finding the site: From Roundup drive south on US 87. The roadcuts begin about 2 miles south of Roundup and extend for 15 miles, with the cuts farthest south from Roundup often producing the best fossils.

Rockhounding

Roundup was once a coal-mining town, and today the rock hound can benefit from the imprint of coal—literally. This highway leading out of Roundup has wonderfully preserved fossil prints of plants in the layers of the roadcuts. The primary strata are Tertiary sandstones and shales belonging to the Fort Union Formation. These layers are easily identifiable, and the best fossil samples are obtained from splitting the layers with a hammer and chisel. Prints of plants can be collected easily above and below the coal vein but are not as finely detailed as prints from splitting the other layers. As with any delicate plant fossil, bring proper packaging equipment for the fragile prints.

Sites 69–73

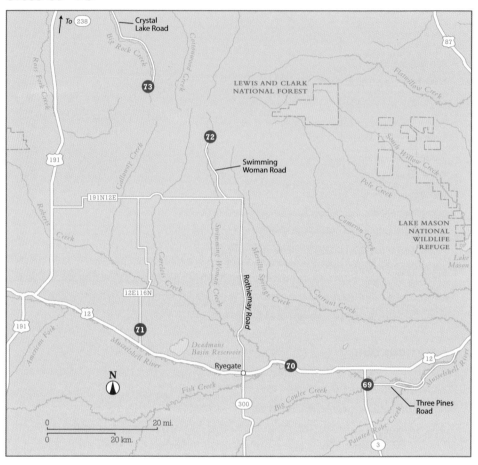

69. Lavina Fossils

The 3-foot-thick deposit of Lavina oyster fossils

See map on page 160.
Land type: Hills, roadcut
GPS: N46° 16.68667' / W108° 55.82167'
Best season: Any
Land manager: Golden Valley County
Material: Fossils
Tools: None
Vehicle: Any
Accommodations: Camping, RV parking, and motels in Roundup
Special attractions: Musselshell Valley Historical Museum in Roundup has a small selection of fossils as well as historical relics.

Finding the site: From the junction of US 12 and MT 3 in Lavina, drive south on MT 3 for 1.8 miles. Turn left onto Three Pines Road just after Cache Creek. Drive 0.3 mile to the roadcut on your right. The northwest portion of the outcrop has the thick layer of fossilized oysters that can be found beneath the dark lignite.

Rockhounding

Although this is a very small site, the 3-foot-thick chunky layer of solid fossilized Cretaceous oysters is proof that good things can come in small roadcuts. This site provides plenty of collecting material and demonstrates the formation of theses rocks in a shore environment with alternating periods of terrestrial plant material and aquatic life. The lignite is a result of the drier times where terrestrial plants flourished, as lignite is a poor grade of coal and forms from these conditions. The alternating layers in the roadcut of sandstone, shale, oyster conglomerate, and lignite are evidence of the changing environments of the past. The oysters found here belong to a single genus and species, *Crassostrea subtrigonalis.*

70. Ryegate Fossils

See map on page 160.

Land type: Hills, roadcut

GPS: N46° 18.73333' / W109° 07.72833'

Best season: Spring through fall

Land manager: Montana DOT

Material: Fossils

Tools: None

Vehicle: Any

Accommodations: Camping, RV parking, and motels in Roundup

Special attractions: Musselshell Valley Historical Museum in Roundup has a small selection of fossils as well as historical relics.

A fossil oyster from Ryegate

Finding the site: From the intersection of US 12 and MT 300 in Ryegate, drive east on US 12 for 6.5 miles to a roadcut on the north side of the road. Search the roadcut for loose oysters or conglomerate rocks. The heaviest concentration observed was at the east end of the roadcut.

Rockhounding

Similar to the fossils found at Lavina, Cretaceous oysters at this particular site can be found easily in the roadcut, although the fossils here are not as concentrated and the lignite layer is not found. The samples are easy to collect and most often complete. A rock hammer is not necessary because the fossils are loose.

71. Shawmut Trace Fossils

See map on page 160.
Land type: Hills, roadcut
GPS: N46° 22.47500' / W109° 31.73833'
Best season: Spring through fall
Land manager: Montana DOT
Material: Fossils
Tools: None
Vehicle: Any
Accommodations: Camping, RV parking, and motels in Roundup
Special attractions: Musselshell Valley Historical Museum in Roundup has a small selection of fossils as well as historical relics.

Finding the site: From Shawmut, 13 miles east of Ryegate on Why 3, head north on FR12E116N and drive 2 miles to the roadcut on your right.

Rockhounding

These fossil burrows and casts of plants are found in the coarse-grained gray sandstone. Keep an eye out for the roadcut with a solid sandstone cap. There are plenty of loose pieces so explore the float along the base of the outcrops.

72. Swimming Woman Creek Fossils

Swimming Woman Creek marine fossils

See map on page 160.
Land type: Mountains
GPS: N46° 42.90667' / W109° 20.79333'
Best season: Spring through fall
Land manager: USFS, Lewis and Clark National Forest
Material: Fossils
Tools: None
Vehicle: Any
Accommodations: Camping and RV parking in Lewis and Clark National Forest; camping, RV parking, and motels in Lewistown
Special attractions: The Central Montana Museum in Lewistown has displays on Montana history and a small collection of fossils and minerals.

Finding the site: From Judith Gap take 279 about 17 miles to Judith Gap

Road/191N12E and turn left. Take Judith Gap Road for 7.8 miles and turn left onto Swimming Woman Road. Drive 9 miles to the parking area. From there follow the forest road into the canyon until you reach the limestone cliffs. You might have to walk in during spring melt, which lasts through late June, but the beautiful drive is worth it.

Rockhounding

The area in and around the Big Snowy Mountains contains fossils in the many outcrops of varying ages from the Early Paleozoic to Mesozoic. The Mississippian Madison Limestone is well exposed, occurs throughout the range, and is said to have a strong presence of various marine fossils such as brachiopods, corals, and crinoids, where it is found. Relatively complete crinoids have also been found in some of the thin-bedded Madison limestone. They are difficult to find and are obviously highly prized. The foothills of the range consist of upturned layers of Mesozoic rocks and marine fossils, such as belemnites, oysters, and crinoids. The Jurassic shales appear to be the best fossil-producing rock in the Mesozoic sequence, but good exposures are not common. There are many roads throughout the range that offer an excellent chance to explore this beautiful area a little longer.

73. Crystal Lake Fossils

See map on page 160.
Land type: Mountains, roadcut, lakeshore
GPS: N46° 48.25440' / W109° 30.69900'
Best season: Summer
Land manager: USFS, Lewis and Clark National Forest
Material: Fossils
Tools: Rock hammer
Vehicle: Any
Accommodations: Camping in Lewis and Clark National Forest; camping, RV parking, and motels in Lewistown
Special attractions: The Central Montana Museum in Lewistown has displays on Montana history and a small collection of fossils and minerals.

Finding the site: From Lewistown drive 8.5 miles west on MT 200 and turn left onto Crystal Lake Road. Follow the signs to Crystal Lake; it is about a 20-mile drive on well-maintained gravel roads. This road has several roadcuts with promising leads; don't limit exploring to just around the lake.

Rockhounding

The area in and around the Big Snowy Mountains contains fossils in the many outcrops of varying ages from the Early Paleozoic to Mesozoic. The Mississippian Madison Limestone is well exposed, occurs throughout the range, and is said to have a strong presence of various marine fossils such as brachiopods, corals, and crinoids, where it is found. Relatively complete crinoids have also been found in some of the thin-bedded Madison limestone. They are difficult to find and are obviously highly prized.

The foothills of the range consist of upturned layers of Mesozoic rocks and marine fossils, such as belemnites, oysters, and crinoids. The Jurassic shales appear to be the best fossil-producing rock in the Mesozoic sequence, but good exposures are not common. There are many roads throughout the range that offer an excellent chance to explore this beautiful area a little longer.

A good place to start collecting is on the road leading to Crystal Lake, a popular camping area in the northwest part of the Big Snowy Mountains. The scenic lake is reason enough to choose this route. The road passes through upturned units of Paleozoic rocks and provides easy access to varying ages of rocks. A search of the roadcuts and cliffs will reveal the fossiliferous nature of many of these beds. More so, however, the limestone gravels of the shore of Crystal Lake produce nice samples of small marine fossils. As you walk farther from shore, larger limestone boulders occur with more variety of fossils. This material is extremely hard to break, and the most efficient collecting is done by picking up pieces that have already been split from the larger rocks.

Site 74

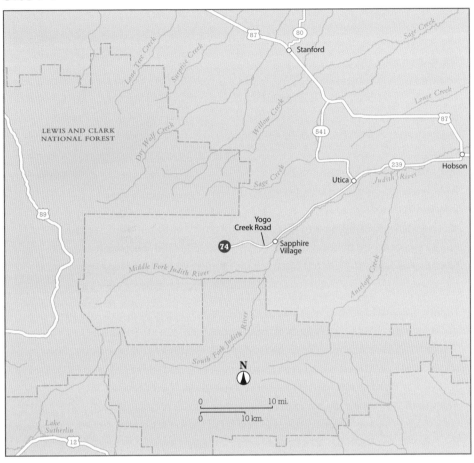

74. Yogo Gold and Fossils

See map on page 167.

Land type: Mountains, stream

GPS: N46° 52.84500' / W110° 21.28000'

Best season: Summer

Land manager: USFS, Lewis and Clark National Forest

Material: Fossils, gold, sapphires

Tools: Rock hammer, gold pan, trowel, fleck flask

Vehicle: High clearance, 4WD

Yogo fossil

Accommodations: Camping and RV parking on-site in Lewis and Clark National Forest; camping, RV parking, and motels in Lewistown and Harlowton

Special attractions: The Central Montana Museum in Lewistown has displays on Montana history and a small collection of fossils and minerals.

Finding the site: These directions are to one particular site, but there are several good panning sites and roadcuts of fossiliferous limestone in the area.

From Sapphire Village, a small town off MT 239 southwest of Lewistown, head south on MT 239/Deadman's Creek Road for 0.4 mile and turn right onto FR 266/Yogo Creek Road. Continue 1.3 miles and stay right at the Y. Drive another 5.3 miles to the limestone cliffs and 0.5 mile farther for the panning area. Adventurous back roaders can continue for another several miles to Elk Creek, 1.4 miles past the Yogo Morris Trailhead, where a very rich site exists for marine fossils. Here you drive up Elk Creek where the road and creek become one and the same. Look for marine fossils and igneous minerals around the Elk Creek sign and in the surrounding washes and outcrops.

Rockhounding

The first thing you will probably wonder is why this isn't a sapphire site. Of all the famed gems in Montana, the Yogo sapphire from the Little Belt Mountains is the most precious. The deep cornflower-blue color of the sapphires found here is world famous. The sapphires of Yogo Creek are the only ones known in the world that naturally occur in a deep uninterrupted blue.

Yogo Gulch was originally a source of placer gold, and it was strictly by chance that the Yogo sapphire was discovered. Because of their relatively high

Searching for fossils near Elk Creek

density, the unusual blue stones were common in the gold-bearing stream gravels. Curiosity resulted in an evaluation of the stones, and then placer operations were begun to retrieve sapphires along with gold. These operations continued up to the investigation and eventual mining of a large 4-mile-long igneous dike containing the sapphires. Today this sapphire mine is the only one in the world that mines the precious stones out of hard rock. Inaccessible to collectors, the mine alternates between activity and closure, and there is a great deal of confidence that gems will be produced from this area for some time to come.

Fortunately, there are still precious treasures for the rock hound to collect in the public land along Yogo Creek. Decent amounts of gold can be found with diligent panning, and reports of sapphires in the pans are common.

If the minerals don't "pan out," there is a great deal of fossiliferous Mississippian limestone exposed throughout the Little Belt Mountains. The approximately 300-million-year-old rocks contain marine fossils that resulted from an ancient shallow sea that once covered the area. The samples to be found in the roadcuts around Yogo Creek are small marine fossils in good condition. Unfortunately, and quite typical of limestone, the rock is difficult to break out of the cut. Your best luck will be had by searching already-loose rocks on the ground. If you have a forest map and a good sense of adventure, it is documented that a highly fossiliferous reef of limestone is located on the south side of Bandbox Mountain just a few miles north of this site and a rich deposit of marine fossils is located along Elk Creek.

Sites 75 and 76

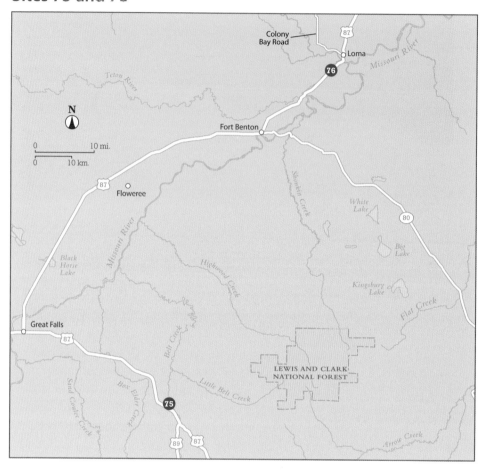

75. Belt Fossils

Belt fossils

See map on page 170.
Land type: High hills, roadcut
GPS: N47° 22.77333' / W110° 55.41333'
Best season: Spring through fall
Land manager: Montana DOT
Material: Fossils
Tools: Rock hammer, chisel
Vehicle: Any
Accommodations: Camping and RV parking in Lewis and Clark National Forest; camping, RV parking, and motels in Great Falls
Special attractions: The C. M. Russell Museum in Great Falls displays art by the cowboy artist along with exhibits on the American West. The Lewis and Clark National Historical Trail Interpretive Center in Giant Springs Heritage State Park details the history of the Lewis and Clark expedition with special emphasis on the Montana leg of the journey. The Giant Springs Visitor Center and Fish Hatchery just outside of Great Falls is located on one of the largest freshwater springs in the world.

Finding the site: From Armington Junction (87/89/200) south of Belt, take Armington Junction Road to an unlabeled road that leads to Belt from the southeast. Note that this sign is not labeled from the north or west of Armington Junction. Follow this road for 2.7 miles northwest to the roadcut on the right. Watch for speeding traffic and park well off the road.

Rockhounding

Jurassic coal seams around the area of Belt tell that once a more tropical time existed in the area that is now Montana. The seams resulted from thick deposits of organic matter in marshy low-oxygen environments that became buried by sand and mud deposited above and below them. Over time the material became coal seams. In 1893 Belt became home to the first commercial coal mine in Montana, and mining of the many coal deposits continued until the 1950s. Today, evidence of the changing environment can be collected around the area of Belt. Leaf and other plant impressions can be found in the layers of shale. The best samples are collected by splitting the dark-gray shale of the roadcut using a hammer and chisel. It could take some time to find a complete sample but, due to the detailed quality of the impressions, it may be worth the energy.

76. Loma Fossils

See map on page 170.
Land type: Hills
GPS: N47° 55.06333' / W110° 32.15667'
Best season: Spring through fall
Land manager: Bureau of Land Management
Material: Fossils, calcite crystals, barite crystals
Tools: None
Vehicle: Any
Accommodations: Camping, RV parking, and motels in Great Falls

Loma baculites fossil

Searching for Loma fossils

Special attractions: Museum of the Northern Great Plains and Museum of the Upper Missouri River in Fort Benton have collaborative exhibits on the history of the area. The Upper Missouri National Wild and Scenic River Visitor Center has information on the wildlife and natural history of the river.

Finding the site: This site is rather small but the gravel road leading into it is full of fossiliferous outcrops so that is an option as well. From the small town of Loma, just north of Fort Benton on US 87, drive to the north end of Loma and turn left onto Colony Bay Road (fossiliferous roadcuts line the road). Drive about 1.5 miles down the road and turn right onto a small dirt BLM road. The first hill is on the right just 0.1 mile down the road, and the second is just 0.2 mile down the road. If the gate is closed, park off Colony Bay Road and walk into the BLM land by following the faint road. A BLM map is recommended.

Rockhounding

Marine shales and sandstones of the Late Cretaceous Colorado Group appear predominately to the north and west of Great Falls, and prized fossil specimens from these rocks can be found, especially along the drainage basins of the Sun, Teton, and Marias Rivers. This site can produce some wonderful specimens of marine fossils, especially ammonites, naculites, and oysters—all of which retain their original mother-of-pearl shell. The fossils are found on either side of the BLM road. The best specimens are found by easily splitting the eroding gray shale and loose shale concretions. Along with the marine fossils, nice barite and calcite crystals can also be found in the shale, some of which replaces fossils.

Site 77

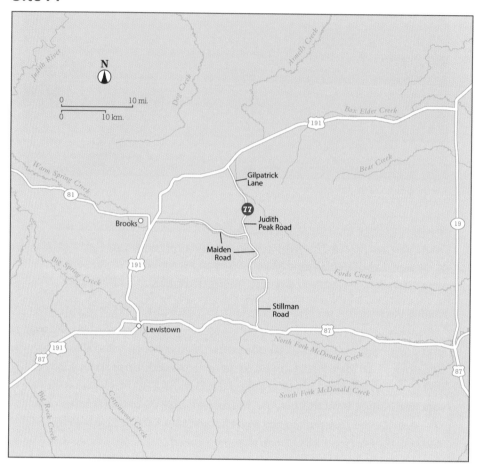

77. Judith Peak Crystals

Judith Peak "Montana Diamond"

See map on page 174.
Land type: Mountains
GPS: N47° 13.06500' / W109° 12.90333'
Best season: Spring through fall
Land manager: Bureau of Land Management
Material: Quartz crystals
Tools: None
Vehicle: Any
Accommodations: Open camping on BLM property; camping, RV parking, and motels in Lewistown
Special attractions: The Central Montana Museum in Lewistown has displays on Montana history and a small collection of fossils and minerals.

Finding the site: From Lewistown drive north on US 191 for 10 miles and turn right onto Warm Springs Canyon Road, following the signs toward the ghost town of Maiden. Drive 9.2 miles and turn left, following the signs to Judith Peak. Continue 3.1 miles to the first good outcrop of a light feldspar rock. You can begin searching at this point. Another good collecting area is located down the faint road on the right; follow it to the dead end, which is a couple of hundred feet after the turnoff. At this location an abundance of crystals were found in the soil at the base of the outcrop. If you continue about 1 mile up the road to Judith Peak, you will reach a radar tower. This area offers a stunning view and the same outcropping with Montana Diamonds, along with large veins of smoky quartz and the occasional crystal pocket. The summit of the peak was at one time the site of a US Air Force radar station. The view from the old radar tower is well worth the short drive. On a clear day most of the nearby mountain systems can be viewed, as well as several interesting geologic features on the prairie to the east.

Rockhounding

Judith Peak is located at 5,808 feet and towers above the valley below. This site is both an amazing place to visit with a spectacular valley view and an excellent site for collecting one of Montana's unique rockhounding treasures. Small

The doubly terminated quartz crystals have a naturally occurring diamond shape and are located in the light porphyritic rock in the roadcuts.

doubly terminated crystals of quartz often referred to as "Montana Diamonds" weather from the porphyritic rock containing them; these are sought by mineral collectors primarily because of their extremely uniform shape. Although interesting and very abundant, they lack the quality of the famed Herkimer "diamonds" of New York and New Jersey. The crystals are not clear but are generally a very dark smoky and opaque variety of quartz. Sizes range from about ¼ to 1 inch in diameter. The crystals have been used as jewelry items by some rock hounds because of their perfect symmetrical form. The porphyry is highly weathered, and the doubly terminated crystals are lying free in the soil and weathered rock in the large roadcuts near the summit of the peak.

The dark quartz crystals are easily seen in the solid rock that composes the roadcut, but they normally cannot be removed from the parent rock without being broken. The texture of the rock containing the crystals demonstrates its plutonic or intrusive nature. The best crystals are those that have weathered naturally from the rock and can be found loose in the gravels, but be selective because most are not of high quality. On any given day it is easy to go home with several dozen crystals found loose in the gravel. The sun offers great assistance to finding the quartz in the soil, as they are more easily seen glistening in the sunlight.

Sites 78 and 79

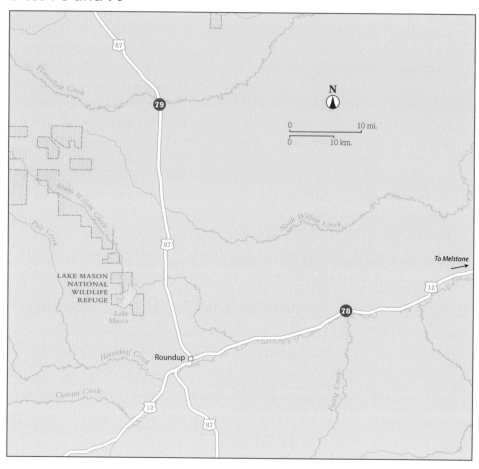

78. Musselshell Fossils

Fine detail of Musselshell fossil plants

See map on page 178.
Land type: Low mountains, roadcut
GPS: N46° 30.73167' / W108° 14.41167'
Best season: Spring through fall
Land manager: Montana DOT
Material: Fossils
Tools: None
Vehicle: Any
Accommodations: Camping, RV parking, and motels in Roundup
Special attractions: Musselshell Valley Historical Museum in Roundup has a small selection of fossils as well as historical relics.

Finding the site: This site consists of several miles of fossiliferous roadcuts between Roundup and Melstone. From Roundup drive east on US 12 following roadcuts of sandstone and shale to Melstone; the best location seems to be directly between the towns beginning about 15 miles out of Roundup.

Rockhounding

Searching through these roadcuts is just as enjoyable as the scenic eastern Montana drive along this highway parallel to the Musselshell River. From Roundup you head east into the oil-pumping fields of Melstone and Ingomar, with, of course, the veins of coal leading the way. Much like the site south of Roundup, carbon prints of trees, branches, leaves, and other organics can be found easily by splitting the sandstone and shale in the roadcuts. Bring suitable packaging materials, as the prints can easily rub off.

79. Flatwillow Creek Fossils

See map on page 178.
Land type: Hills, roadcut
GPS: N46° 46.804' / W108° 36.315'
Best season: Spring through fall
Land manager: Montana DOT
Material: Fossils, gypsum, selenite
Tools: Rock hammer
Vehicle: Any
Accommodations: Camping, RV parking, and motels in Roundup
Special attractions: Musselshell Valley Historical Museum in Roundup has a small selection of fossils as well as historical relics.

Finding the site: From Roundup drive 24 miles north on US 87 to several rock outcrops along the road. Collecting areas are located in a large spread on both sides of US 87, about 3 to 5 miles past the MT 244 junction to Winnett.

Rockhounding

This area can produce some interesting fossils. The first locality is encountered about 3 miles north of the MT 244 junction to Winnett. Here, as US 87 descends into the valley of Flatwillow Creek, several roadcuts in exposures of the Cretaceous Colorado Shale are found. The Mosby Sandstone is well exposed in the northernmost roadcut on the east side of the highway. Specimens of fossil oysters can be picked up on the surface roadcut, while

Fossils and gypsum at Flatwillow Creek

sandy concretions containing a fair abundance of fossil snails are found here and there. Some of the adjacent roadcuts contain limestone concretions with attractive white and brown calcite crystals. Crystals of selenite are abundant on the surface of all the roadcuts. About 2 miles farther north on US 87 (5 miles north of the MT 244 junction to Winnett) are some low roadcuts on both sides of the highway exposing the Cretaceous Mowry Shale. By splitting the light-colored, thin-bedded layers, one can find abundant dark-brown fish scales that average ¼ inch in size. Some magnification (10X hand lens) will reveal the concentric ring pattern on the surface of the scales.

Emilla Palenlus searches the fossil deposit.

Sites 80–84

80. Little Rockies Gold

See map on page 182.
Land type: Mountains, riverbed
GPS: See directions below
Best season: Late spring through fall
Land manager: Private
Material: Gold, garnets
Tools: Gold pan, trowel, flask
Vehicle: High-clearance
Accommodations: Camping and RV parking in Little Rocky Mountains; small motel in Zortman; camping, RV parking, and motels in Malta
Special attractions: Fossil Hill, which is full of belemnites, oysters, small ammonites, and calcite crystals. Locals will help you acquire access to this private land.

Finding the site: Just northeast of the junction of MT 66 with US 191 (between Malta and Lewistown), turn west onto the road leading to Zortman. Drive 7 miles, staying left and following the signs to Zortman. The road that leads you to town will take you straight to the Zortman Garage and Motel, which also hosts a small rock shop that sells gold pans, tools, and various supplies. Inquire within the motel office about land status and locations.

Rockhounding

Placer gold was first discovered in this area in 1884, but ten years of continuous placer operations took place before the lode deposits containing the gold were mined. Apparently, the alluvial deposits in which the gold was first discovered are very extensive, but the quantity of water needed to work these deposits is not available. Lode deposits have since provided the major amount of recoverable gold in the district. Several mines near Zortman and Landusky have been reactivated and there are many claims in the area.

The unusual occurrence of the Little Rockies, a small rugged mountain mass in the plains of northern Montana, is attributed to volcanism. The Little Rockies are basically interpreted as a laccolithic structure, that is, an area domed upward as a result of molten magma intruding between sedimentary rock layers and pushing them from below. A later period of erosion has exposed the igneous rock core and the metalliferous veins it contains, with gold and silver being the main metals of interest.

81. Fossil Hill

See map on page 182.

Land type: Mountains, roadcut

GPS: N47° 54.67140' / W108° 23.81880'

Best season: Any

Land manager: Phillips County

Material: Fossils

Tools: None

Vehicle: Any

Accommodations: Camping and RV parking in Little Rocky Mountain, small motel in Zortman; camping, RV parking, and motels in Malta

Special attractions: Gold panning in Zortman. The Zortman Garage and Motel has a rock shop that sells equipment and supplies. Also inquire there on the status of Fossil Hill, an area behind the Zortman dump that is chock-full of belemnites, oysters, small ammonites, and calcite crystals. This site is popular with universities, who have conducted digs that have uncovered important dinosaur fossils on several occasions.

Finding the site: The first roadcut is on the way to Zortman. Just northeast of the junction of MT 66 with US 191 (between Malta and Lewistown), turn west onto the road leading to Zortman. Drive about 5 miles, park, and explore the limestone roadcuts for oysters, belemnites, and other small marine fossils.

Rockhounding

This location can satisfy the paleontology rock hound as well as someone who has the "gold fever." The best collecting sites are on private property, and locals are helpful in acquiring access to Fossil Hill. The key to collecting in these locations is familiarization with the Madison limestone. It's not hard to locate the light-colored rocks that surround the mountain range. Paleozoic and Mesozoic rock layers are turned up sharply along the margins of the Little Rockies, and exposures of various fossiliferous formations often occur in the area. The marine Jurassic shales here continue to yield relatively common specimens of the oyster Gryphaea, the belemnite Pachyteuthis, the star-shaped crinoids Pentacrinus, as well as less-common species of ammonites and brachiopods.

82. Malta Dinosaur Digs

See map on page 182.

Land type: Badlands

GPS: N 48° 21.64836' / W 107° 52.05918'

Best season: Spring through fall

Land manager: Great Plains Dinosaur Museum

Material: Fossils

Tools: Provided

Vehicle: Any

Accommodations: Camping, RV parking, and motels in Malta

Special attractions: The Great Plains Dinosaur Museum has impressive paleontology displays and dinosaur skeletons.

Finding the site: This museum is located at 405 North 1st St. East, Malta, MT 59538.

Rockhounding

The dinosaur country around Malta has been home to some incredible discoveries, including "Leonardo" the "mummy" dinosaur. The skeleton of Leonardo is so complete it includes fossil skin, muscles, and upon a giant CAT scan, a heart, tongue, and stomach contents. The formations around Malta are littered with dinosaur fossils, and paleontologists have a hard time keeping up with the erosion. The Great Plains Dinosaur Museum is one of the better small museums in the state and operates as a nonprofit that helps to incorporate the public into the academic dinosaur experience. All finds become part of the museum or a collection and they offer a variety of programs for different ages. You can book tours as short as a day (for around $150 at the time of publication). All profits go to the museum. For more information go to www .greatplainsdinosaurs.org.

83. Malta Fossils

Searching for fossils

See map on page 182.
Land type: High hills
GPS: N48° 35.808' / W107° 43.944'
Best season: Spring through fall
Land manager: Montana DOT
Material: Fossils, calcite crystals
Tools: Rock hammer
Vehicle: Any
Accommodations: Camping, RV parking, and motels in Malta
Special attractions: The Great Plains Dinosaur Museum has impressive paleontology displays and dinosaur skeletons.

Finding the site: From Malta, drive north on US 191 for 19.5 miles to a large defined roadcut and turnout where the material is located.

Rockhounding
The Cretaceous Pierre Shale, a dark-gray and brown shale north of Malta, occurs in a very small section along the highway, but it's abundant in ammonites. The fossils, along with calcite crystals, are easily found by splitting the large shale concretions found on and around the hillsides.

84. Glasgow Chert

See map on page 182.
Land type: Hills
GPS: N48° 13.39100' / W106° 37.81400'
Best season: Spring through fall
Land manager: Bureau of Land Management
Material: Jasper, chert, chalcedony
Tools: None
Vehicle: Any
Accommodations: Camping, RV parking, and motels in Fort Peck and Glasgow

Glasgow chert

Special attractions: Fort Peck Dam Interpretive Center and Museum has displays on Fort Peck Dam as well as geology, with an impressive paleontology exhibit and life-size T. rex models. Fort Peck Dam is one of the largest earth-filled dams in the world.

Finding the site: From the east end of Glasgow on US 2, turn north onto Skylark Road and follow the signs to the right 1 mile to the OHV area. Park at the restroom and explore the hilltops.

Rockhounding

The whipping eastern Montana wind has naturally polished many of the siliceous rocks you can find on the exposed hilltops around the Glasgow area. At this site you can easily collect the chert, jasper, and chalcedony. The colors are predominantly black, brown, orange, and yellow. Many don't need polishing, but a tumble would bring out the brilliant colors.

Sites 85 and 86

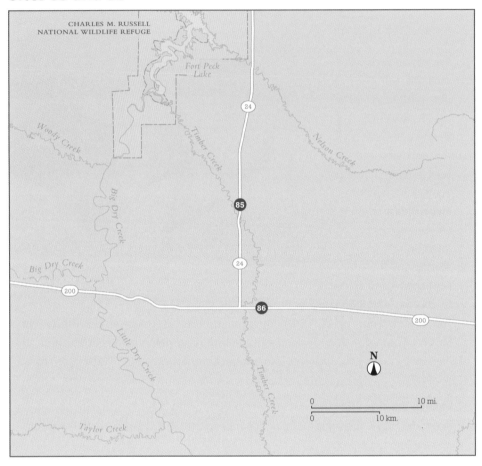

85. Timber Creek Fossils

Selenite crystals from Timber Creek

See map on page 188.
Land type: Hills
GPS: N47° 24.977' / W106° 10.242'
Best season: Spring through fall
Land manager: Bureau of Land Management
Material: Fossils, calcite crystals, petrified wood, selenite, iron concretions
Tools: None
Vehicle: Any
Accommodations: Camping and RV parking on Fort Peck; camping, RV parking, and motels in Glasgow
Special attractions: Fort Peck Dam Interpretive Center and Museum has displays on Fort Peck Dam as well as geology, with an impressive paleontology exhibit and life-size T. rex models. Fort Peck Dam is one of the largest earth-filled dams in the world.

Finding the site: Numerous fossiliferous roadcuts line the road between Fort Peck and MT 200. It is illegal to collect within the C. M. Russell preserve so be sure to stick to roadcuts on BLM land or outside the preserve. From the junction of MT 24 and MT 200 about halfway between Circle and Jordan, head north on MT 24 for 6.2 miles to the first roadcut along the east side of the road. The roadcuts extend about 30 miles and there are several options for rockhounding off the highway on BLM land, so consider getting a BLM map of the region. Stop collecting when you enter the preserve.

Rockhounding

Several barriers keep the rock hound from collecting on Fort Peck Lake, home to some of the best marine fossil specimens to be found in the state. The main reason is that vertebrate collecting of any kind is illegal by federal law, and the law is firmly enforced within the Fort Peck area. Collecting rocks of any kind is prohibited within the Charles M. Russell National Wildlife Refuge, and there are signs posted everywhere warning that violators will be prosecuted and can face jail time for disobeying. One might think this is to preserve the refuge in its natural state, but a rock hound may look at it as a sign of how spectacular the fossils are that wash up onshore. The shoreline of this area, particularly at Bear Creek, is intermingled with eroded Cretaceous Bearpaw Shale, and all the mollusks can be found along the shoreline loose in the silt or in the rocks along the lakeshore.

The area around Fort Peck Lake provides some of the best exposures of Cretaceous Bearpaw Shale, Judith River, and Hell Creek Formations to be found in the state. In the early 1900s Barnum Brown, a paleontologist for the American Museum of Natural History, discovered in the 65-million-year-old Cretaceous sediments of the Hell Creek Formation the skeletal remains of a remarkable creature: the Tyrannosaurus rex, which has come to epitomize the dinosaur. Skeletons of the armored, three-horned dinosaur triceratops are also frequently found here, as well as the remains of hadrosaurs and the rare ankylosaurs.

The sediments of the Hell Creek Formation southeast of Fort Peck Lake have produced more than just dinosaurs. Numerous skeletal remains of tiny mammals have been collected from several badlands areas as well. Unfortunately it is illegal to collect anything within the preserve. The directions for this location lead to roadcuts outside the preserve, which are legal to collect on. Remember that vertebrates (anything with a backbone) are always illegal to collect on public land. This is not an issue, however, since the selenite and plant fossils are so impressive.

86. Jordan Fossils

See map on page 188.
Land type: Hills
GPS: N47° 19.53667' / W106° 08.52000'
Best season: Spring through fall
Land manager: Montana DOT
Material: Fossils
Tools: None
Vehicle: Any
Accommodations: Camping, RV parking, and motels in Jordan
Special attractions: The Garfield County Museum has excellent paleontology displays including a locally excavated triceratops skeleton.

Jordan plant fossils

Finding the site: Located a couple miles east of the rest area at mileposts 244 to 250 on MT 200, 37 miles east of Jordan. GPS is taken at milepost 250, east of Skull Creek and 1.3 miles east of the rest area at a roadcut on the north side of the road.

Rockhounding

Perhaps the most famous of all deposits in Montana is the Cretaceous Hell Creek Formation. The dinosaur-bearing unit is well known for significant discoveries including the iconic dinosaurs Tyrannosaurus rex and triceratops. The multicolored mudstones and sandstones that the formation is composed of form the classic "badland" topography. Fossils of many creatures, along with plants and a variety of minerals, easily erode out of the crumbling layers. The formation is also famous for the layer of iridium, an extraterrestrial impact dust, from the asteroid that hit Earth and eliminated the dinosaurs. The iridium layer of the Cretaceous-Paleogene boundary can be viewed with the naked eye at only a few places in the world and most are in Garfield County, Montana. Here in this stretch of road, you can see massive outcroppings of the Hell Creek Formation, but most is on private ranch land. Several roadcuts near Jordan do offer a chance to stop and appreciate the Hell Creek Formation. Remember it is illegal to collect vertebrates (bones) so stick to the petrified wood, selenite, and iron concretions.

Site 87

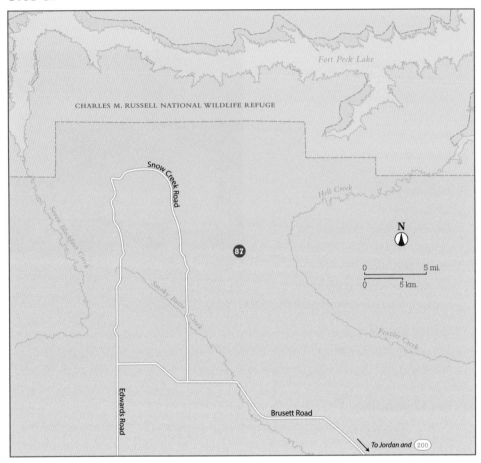

87. PaleoWorld Dinosaur Digs

See map on page 192.
Land type: Badlands
GPS: N47° 30.86580' / W107° 10.26780'
Best season: Summer
Land manager: Private, PaleoWorld
Material: Fossils
Tools: Provided
Vehicle: Any
Accommodations: Provided
Special attractions: The Garfield County Museum has excellent paleontology displays including a locally excavated triceratops skeleton.

Finding the site: PaleoWorld dinosaur digs are usually based out of Jordan, inquire with PaleoWorld for current dig sites and meeting locations.

Rockhounding

PaleoWorld offers an opportunity for the public to join dinosaur digs in the Hell Creek Formation around the Jordan area. Perhaps the most famous of all deposits in Montana is the Cretaceous Hell Creek Formation. The dinosaur-bearing unit is well known for significant discoveries including the iconic dinosaurs Tyrannosaurus rex and triceratops. The multicolored mudstones and sandstones that the formation is composed of form the classic "badland" topography. Fossils of many creatures, along with plants and a variety of minerals, easily erode out of the crumbling layers. The formation is also famous for the layer of iridium, an extraterrestrial impact dust, from the asteroid that hit Earth and eliminated the dinosaurs. The iridium layer of the Cretaceous-Paleogene boundary can be viewed with the naked eye at only a few places in the world and most are in Garfield County, Montana. With this company you can join a dig for a day or longer. In 2015 the prices per day were $160, with housing at local motels, for day trips and trailers at a base camp. The location and material of each dinosaur excavation varies, so contact PaleoWorld for more information at paleoworld.org.

Site 88

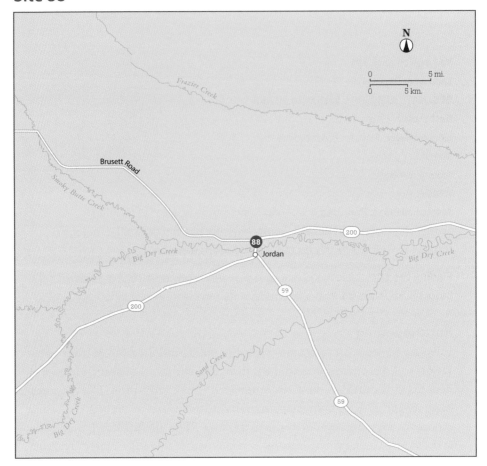

88. Paleotrek Dinosaur Digs

See map on page 194.

Land type: Badlands

GPS: N47° 19.33667' / W106° 54.60000'

Best season: Limited summer dates

Land manager: Private, Paleotrek

Material: Fossils

Tools: Provided

Vehicle: Any

Accommodations: Camping, RV parking, and motels in Jordan and Billings

Special attractions: The Garfield County Museum has excellent paleontology displays including a locally excavated triceratops skeleton.

Finding the site: Paleotrek dinosaur digs are based out of the Jordan area. Contact Paleotrek for more information.

Rockhounding

Paleotrek offers an opportunity for rock hounds with prior dinosaur excavation and prospecting experience to join professional dinosaur digs in the Hell Creek Formation around the Jordan area. The organization has limited dates and is academically operated through a collaboration with St. Louis Science Center, Eastern Missouri Society for Paleontology. Perhaps the most famous of all deposits in Montana is the Cretaceous Hell Creek Formation. The dinosaur-bearing unit is well known for significant discoveries including the iconic dinosaurs Tyrannosaurus rex and triceratops. The multicolored mudstones and sandstones that the formation is composed of form the classic "badland" topography. Fossils of many creatures, along with plants and a variety of minerals, easily erode out of the crumbling layers. The formation is also famous for the layer of iridium, an extraterrestrial impact dust, from the asteroid that hit Earth and eliminated the dinosaurs. The iridium layer of the Cretaceous-Paleogene boundary can be viewed with the naked eye at only a few places in the world and most are in Garfield County, Montana. Through Paleotrek, degreed paleontologists offer programs for a variety of ages and backgrounds. In 2015 the fee was $50/day with a two-day minimum. For more information visit paleotrek.com.

Site 89

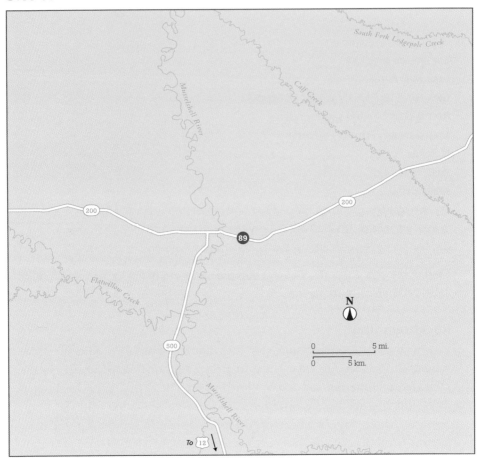

89. Winnett Fossils

See map on page 196.
Land type: Badlands
GPS: N46° 59.35667' / W107° 51.60833'
Best season: Spring through fall
Land manager: Montana DOT
Material: Fossils
Tools: None
Vehicle: Any
Accommodations: Camping, RV parking and motels in Jordan
Special attractions: The Garfield County Museum has excellent paleontology displays including a locally excavated triceratops skeleton.

Finding the site: From the rest stop on MT 200, head east to mile marker 158. The roadcuts of dark-gray shale continue to Winnett.

Rockhounding

Cretaceous marine fossils and calcite crystals can be found along MT 200 from Winnett to Jordan. Look for the large dark-gray eroding shales. The fossils easily weather out.

Sites 90 and 91

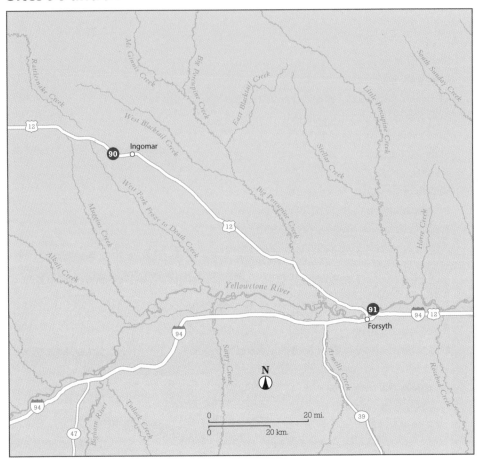

90. Ingomar Fossils

See map on page 198.
Land type: Hills, roadcut
GPS: N46° 35.20000' / W107° 26.07667'
Best season: Spring through fall
Land manager: Montana DOT
Material: Fossils, petrified wood, iron concretions
Tools: None
Vehicle: Any
Accommodations: Camping, RV parking, and motels in Roundup
Special attractions: None

Finding the site: From US 12 in Ingomar, head west about a mile to the roadcuts along US 12. The GPS and specimens collected are from the roadcut at mile 227.

Rockhounding

You have an opportunity to sift through the Hell Creek Formation here in Ingomar. Perhaps the most famous of all deposits in Montana is the Cretaceous Hell Creek Formation. The dinosaur-bearing unit is well known for significant discoveries including the iconic dinosaurs Tyrannosaurus rex and triceratops. The multicolored mudstones and sandstones that the formation is composed of form the classic "badland" topography. Fossils of many creatures, along with plants and a variety of minerals, easily erode out of the crumbling layers. The formation is also famous for the layer of iridium, an extraterrestrial impact dust, from the asteroid that hit Earth and eliminated the dinosaurs. The iridium layer of the Cretaceous-Paleogene boundary can be viewed with the naked eye at only a few places in the world and most are in Garfield County, Montana. Here in this stretch of road, you can see massive outcroppings of the Hell Creek Formation, but most is on private ranch land. Several roadcuts near Jordan do offer a chance to stop and appreciate the Hell Creek Formation. Remember it is illegal to collect vertebrates (bones), so stick to the petrified wood, selenite, and iron concretions.

91. Forsyth Agate

See map on page 198.
Land type: River, gravel bar
GPS: N46° 16.46667' / W106° 40.74500'
Best season: Spring through fall
Land manager: Montana Fish, Wildlife, and Parks
Material: Agate, jasper, petrified wood
Tools: None
Vehicle: Any
Accommodations: Camping, RV parking, and motels in Forsyth
Special attractions: None

Finding the site: From I-94 take exit 95 toward Forsyth. Following the signs to the Rosebud Recreation Area, drive north, and turn right onto 15th Avenue. Go 0.4 mile to the recreation area for the Yellowstone River.

Rockhounding

The best and most efficient times to search for agate are in spring, just as the snow melts but the rivers are not yet full, and during late summer, as the river reaches its driest point. In Forsyth search the banks, gravel bars, and shoreline for agate, jasper, and petrified wood. A sunny afternoon is also a good time to look for the transparent agate.

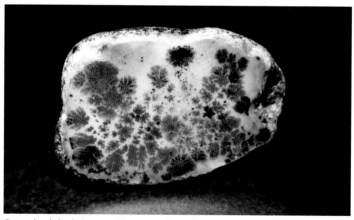

Cut and polished Montana River agate from the Agate Stop

Site 92

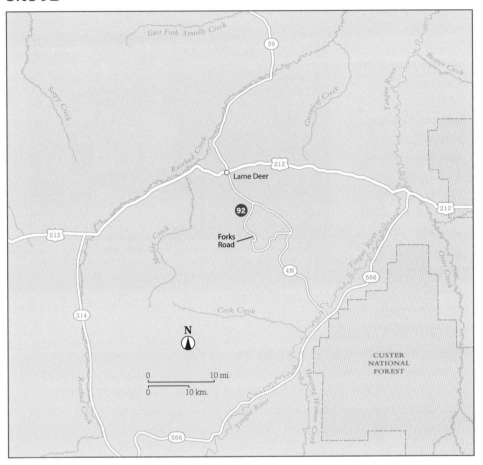

92. Castle Rock Lake Fossils

See map on page 201.
Land type: Hills, lakeshore
GPS: N45° 33.82667' / W106° 38.35000'
Best season: Spring through fall
Land manager: Colstrip Parks and Recreation District
Material: Fossils
Tools: None
Vehicle: Any
Accommodations: Camping, RV parking, and motels in Forsyth
Special attractions: None

Detailed leaf imprints and bark fossils from Castle Rock Lake

Finding the site: From MT 39 in Colstrip, drive to the north end of town just before the mile 24 marker, take the road left into the Castle Rock Lake Recreation Area, and drive 0.4 mile to the parking area. The specimens collected and GPS coordinates were taken to the left in a large pile of red clinker on the shore by the culvert.

Rockhounding
The area around Colstrip is famous for its large deposits of low-sulphur coal. Deposits occur as relatively thick seams in the Tertiary Fort Union Formation and have been strip-mined for many years. Exposures of the Fort Union Formation in the area contain well-preserved fossil leaf prints, similar to those found near Miles City. A good place to search for these imprints is at the Castle Rock Lake Recreation Area. A level hiking trail follows the lakeshore and strides by several small but productive cuts of red clinker ("scoria") that contain fossils. The walk around the small, quiet lake isn't so bad either.

Sites 93–95

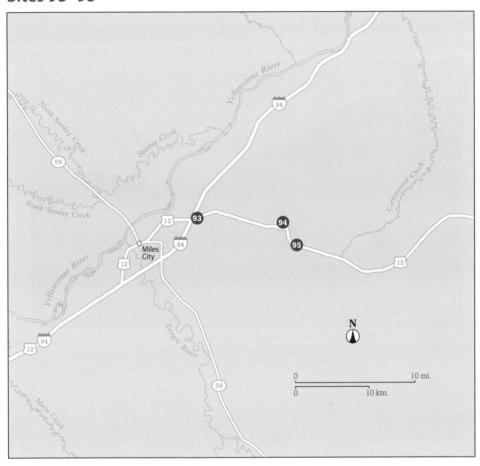

93. Miles City Fossils

Exposures of the Fort Union Formation around Miles City

See map on page 203.
Land type: Hills
GPS: N46° 25.61667' / W105° 46.78833'
Best season: Spring through fall
Land manager: Montana DOT
Material: Fossils, agate, chalcedony
Tools: None
Vehicle: Any
Accommodations: Camping, RV parking and motels in Miles City
Special attractions: The Range Riders Museum and Bert Clark Gun Collection in Miles City features a collection of western antiques, artifacts, and small geology displays.

Finding the site: This site is located just east of the I-94 and US 12 junction in Miles City.

Rockhounding

The Paleocene Fort Union Formation is exposed in the Miles City area and offers prime rockhounding. At this site you will find wood and plant fossils in a medium-grained sandstone as well as iron concretions. Also loose in the alluvium, keep an eye out for agate and chalcedony.

94. Strawberry Hill Petrified Wood

See map on page 203.

Land type: Badlands

GPS: N46° 25.40667' / W105° 40.60500'

Best season: Late spring through summer

Land manager: Bureau of Land Management

Material: Petrified wood, iron concretions

Tools: Rock hammer

Vehicle: High clearance

Accommodations: Camping, RV parking, and motels in Miles City

Special attractions: The Range Riders Museum and Bert Clark Gun Collection in Miles City features a collection of western antiques, artifacts, and small geology displays.

Finding the site: From the junction of I-94 and US 12, take US 12 east toward Baker for 5.6 miles. Turn left onto an unmarked dirt road near a large gravel lot and drive 0.7 mile into an area of badlands. Park and explore.

Rockhounding

Around Miles City in the beautifully colored Fort Union Formation, logs, stumps, and limbs of fossilized sequoia trees periodically are found weathering in the hillsides. The gravels of some of the stream bottoms in the region contain relatively large fragments of colorful agatized petrified wood, ideal for rock gardens. Keep the government regulation concerning petrified wood in mind if you are on public land and decide to collect any of this material. In some cases, one or two pieces could easily fill the yearly quota.

A fossil tree stump eroding at Strawberry Hill

Petrified wood of Strawberry Hill

Sandstone formations high in the badlands can produce wonderful specimens of petrified wood and iron concretions. Search in and around wash beds for petrified wood, which is generally rusty colored and quite large. This is a large site with much to explore—a BLM map is recommended.

95. Fort Union Fossils

See map on page 203.

Land type: Hills, roadcut

GPS: N46° 24.32000' / W105° 39.56000'

Best season: Any

Land manager: Montana DOT

Material: Fossils

Tools: Rock hammer, chisel

Vehicle: Any

Accommodations: Camping, RV parking, and motels in Miles City

Special attractions: The Range Riders Museum and Bert Clark Gun Collection in Miles City features a collection of western antiques, artifacts, and small geology displays.

Finding the site: From the junction of I-94 and US 12, drive east on US 12 for 6.5 miles to a large roadcut of red clinker northwest of mile marker 13.

Rockhounding

The Tertiary Fort Union Formation, which is so predominant in the eastern part of Montana, has been a prolific source of extremely well-preserved fossil leaf imprints. The imprints of hardwood trees including birch, elm, oak, and maple are most common. Most exposures of the Fort Union Formation in highway roadcuts have the potential of producing detailed fossils if the thin-bedded sandstones and hard shales are carefully split with a hammer and chisel. The red clinker beds will disclose fine delicate impressions of leaves. These red beds represent clays and mudstones that were baked by the natural burning of coal layers underground. The highly baked clays that now outcrop along the crests of the hills in the area have the appearance of volcanic rock and are often referred to as "scoria." At this particular site there is plenty of material to be found by searching the piles of loose clinker along the base of the roadcuts and splitting the larger pieces.

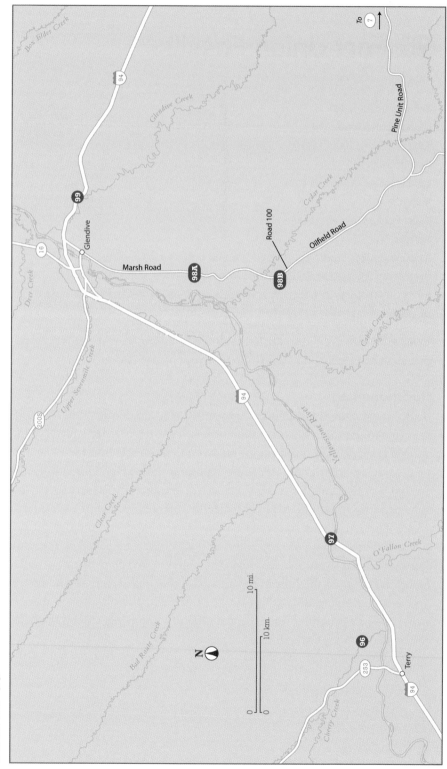

Sites 96–99

96. Terry Agate

Jessica McCartney and her massive Cherry Creek agate

See map on page 208.
Land type: Creek, gravel bars
GPS: N46° 49.55167' / W105° 15.65667'
Best season: Spring through fall
Land manager: Bureau of Land Management
Material: Agate, jasper, petrified wood, chalcedony, chert, fossils
Tools: None
Vehicle: Any
Accommodations: Camping, RV parking, and motels in Glendive
Special attractions: The Prairie County Museum in Terry has small displays on agate and hosts the Evelyn Cameron Gallery, unique photographs taken by one of Montana's early female pioneers.

Cherry Creek agates, petrified wood, jasper, and fossils

Finding the site: From I-94 in Terry take exit 176 to MT 253 north toward the backcountry byway. Drive 2.3 miles and turn right onto Coal Creek Road, a well-graded gravel road. Drive 1.7 miles to unlabeled Cherry Creek. Explore upriver along the public land on the left (northwest) side of the road. Consider taking the backcountry byway after your Cherry Creek rockhounding experience for some amazing badland vistas.

Rockhounding

When the Yellowstone River is too high, this is an excellent site for agates, jasper, petrified wood, fossils, and more. The gravels here are relatively unexplored and extremely rich in sought-after material. Head to the north along the creek to explore the gravel bars and drainages. Excellent-quality translucent agates, patterned jaspers, colorful cherts, fossil imprints in chert, petrified wood, and agatized wood are easily collected at this stop.

97. Fallon Agate

See map on page 208.

Land type: River, gravel bars

GPS: N46° 51.40667' / W105° 06.87667'

Best season: Spring through fall

Land manager: Montana Fish, Wildlife, and Parks

Material: Agate, jasper, chalcedony, petrified wood

Tools: None

Vehicle: Any

Accommodations: Camping, RV parking, and motels in Glendive

Special attractions: Makoshika State Park, a prime area to observe badlands topography

Finding the site: From I-94 near Fallon, take exit 185 to the frontage road on the south side of the interstate. Drive north on the frontage road for 1.5 miles, cross the bridge, and turn right into the boat launch and parking area. Walk to the river and search the gravel bars.

Rockhounding

Like any agate-hunting locality in Montana, the absolute best time of year to collect is in early spring and late summer when the river is at its lowest point. The best locality to collect agate at the Fallon boat launch area is about a 0.25-mile walk north along the river to some large gravel bars.

A decent amount of Montana moss agate, red jasper, and smoothed petrified wood can be found here by walking slowly, preferably on a sunny day, along the gravel bars. There are reports of the popular blue-gray Montana agate being found here as well. Montana Fairburn agate, an agate that has characteristic potato-eye swirls of color, has been found here.

98. Glendive Fossils

Agate, petrified wood, iron concretions, and plant fossils from Glendive

See map on page 208.

Land type: Badlands

GPS: Site A: N46° 59.64000' / W104° 44.52333'; Site B: N46° 54.57000' / W104° 45.02167'

Best season: Spring through fall

Land manager: Bureau of Land Management

Material: Fossils, petrified wood, agate, iron and sandstone concretions, selenite crystals, barite crystals

Tools: Rock hammer

Vehicle: Any

Accommodations: Camping and RV parking in Makoshika State Park; camping, RV parking, and motels in Glendive

Special attractions: Sculptured by running water, Makoshika State Park near Glendive is a spectacular example of badlands topography. The badlands are composed mainly of soft shales and sandstones, and the sediments of the Tertiary Fort Union Formation and the Cretaceous Hell Creek Formation in eastern Montana are easily affected by the agents of weathering and erosion. The remains of fossil plants, dinosaurs, and small mammals, some exceedingly rare, have been removed from these beds. The Makoshika Dinosaur Museum in central Glendive hosts many paleontology displays. The Agate Stop, located on MT 16 in nearby Savage, is by far the most spectacular display and jewelry store for Montana agates in the state.

Finding the site: Collecting anywhere within Makoshika State Park is prohibited and strictly enforced. Some of the more interesting fossils worth searching for outside Makoshika State Park are the leaf prints in the sandstone and shale layers of the Fort Union Formation, and the fossil "figs" and reversed casts of pinecones in the Hell Creek Formation. The sandstone casts called "figs" have the same

The original colors of the bivalve shell from Glendive

general shape as present-day figs, but their actual identity continues to be debated by paleo-botanists. If you stop by the visitor center at the entrance of the park, a map can be requested for the BLM land where collecting is legal.

The first site (site A) is on BLM land and can be reached from Glendive by driving south on Main Street (Road 100/MT 335) for 8.5 miles until the road turns to gravel. At this point there should be a large parking area on both sides of the road predominantly for the Glendive Short Pine OHV Area. Park here and explore hillsides for marine fossils in the shale.

The next site (site B) is reached by driving 8 miles down Road 100 from the previous site. Stay right at the Y where the "Dawson County 120" sign is and the BLM badlands are on the south side of the road for several miles. Search this area for petrified wood, sandstone concretions, fossilized plants, sequoia cones, and leaf casts.

Rockhounding

Site A: At the first site, outcrops of the Late Cretaceous Pierre Shale occur in roadcuts, bluffs, and stream banks along Cedar Creek south of Glendive. Only the uppermost portion of this entire sequence of strata in the Pierre Shale is exposed, so the fauna that can be collected is limited in variety. Among the fossils to be found here are numerous pelecypods, several species of ammonites, many different genera and species of small gastropods, occasional nautiloids, and rare echinoids. Some of the limestone concretions near the top of the dark exposures of Pierre Shale consist almost entirely of pelecypods. Careful breaking of the concretions may uncover excellent specimens of mollusks, many with their original mother–of–pearl shells still intact.

Periodically, septarian concretions are found that contain unusual and attractive deposits of calcite and barite crystals. Lying free on the mounds of gray bentonite that occur throughout the area are abundant "fishtail twin" crystals of selenite and less-common gray nodules composed of radiating barite crystals.

Concretions that have washed down from higher levels litter the creek bottoms, and good fossils often can be found here. Large pieces of agatized wood frequent the gravels of the creek bottoms, but their source is uncertain.

Site B: Exposures of the Cretaceous Fox Hills and Hell Creek Formation directly to the east and west of the Pierre Shale outcrops along Cedar Creek are recognized by the presence of light-colored sandstones and greenish-gray shales. These rock units contain fossils similar to those found in the Makoshika State Park area described earlier. Even though vertebrate fossils ranging from crocodile teeth to whole dinosaurs are found in this area, remember that vertebrate collecting is illegal on public land.

The rock units also contain some unusual reddish sandstone concretions, most likely with high iron contents, which occur as single spheres or groups of spheres. These make interesting additions to a rock collection and are found in gullies and ravines where they have weathered from the rock. Some exposures of the Fox Hills Sandstone just above the outcrops of Pierre Shale will provide specimens of a very unusual fossil. They are rust colored, cylindrical in shape, about 1 inch across, and vary in length up to about 1 foot. They are commonly referred to as "petrified corncobs," probably in reference to their bumpy exterior and general appearance, but they are actually the fossil casts of shrimp burrows. Fossil plants are found in the brownish shale.

99. Baisch's Dinosaur Digs

See map on page 208.
Land type: Badlands
GPS: N47° 06.52333' / W104° 38.07833'
Best season: Spring through fall
Land manager: Private, Baisch's Dinosaur Digs LLC
Material: Fossils

Tools: Provided

Vehicle: Any

Accommodations: Camping, RV parking, and motels in Glendive

Special attractions: Sculptured by running water, Makoshika State Park near Glendive is a spectacular example of badlands topography. The badlands are composed mainly of soft shales and sandstones, and the sediments of the Tertiary Fort Union Formation and the Cretaceous Hell Creek Formation in eastern Montana are easily affected by the agents of weathering and erosion. The remains of fossil plants, dinosaurs, and small mammals, some exceedingly rare, have been removed from these beds. The Makoshika Dinosaur Museum in central Glendive hosts many paleontology displays. The Agate Stop, located on MT 16 in nearby Savage, is by far the most spectacular display and jewelry store for Montana agates in the state.

Finding the site: These digs leave from Glendive. Contact Baisch's Dinosaur Digs for more information at (406) 365-4133 or www.dailydinosaurdigs.com.

Rockhounding

Baisch's is a family-owned company of local ranchers who offer an opportunity for the public to join dinosaur digs in the Hell Creek Formation on their property around the Jordan area. The organization has limited dates and is academically operated through a collaboration with the St. Louis Science Center, Eastern Missouri Society for Paleontology. Perhaps the most famous of all deposits in Montana is the Cretaceous Hell Creek Formation. The dinosaur-bearing unit is well known for significant discoveries including the iconic dinosaurs Tyrannosaurus rex and triceratops. The multicolored mudstones and sandstones that the formation is composed of form the classic "badland" topography. Fossils of many creatures, along with plants and a variety of minerals, easily erode out of the crumbling layers. The formation is also famous for the layer of iridium, an extraterrestrial impact dust, from the asteroid that hit Earth and eliminated the dinosaurs. The iridium layer of the Cretaceous-Paleogene boundary can be viewed with the naked eye only at a few places in the world and most are in Garfield County, Montana. In 2015 the fee was $100 a day per person and excursions were tailored to the ages and backgrounds of the clients.

Site 100

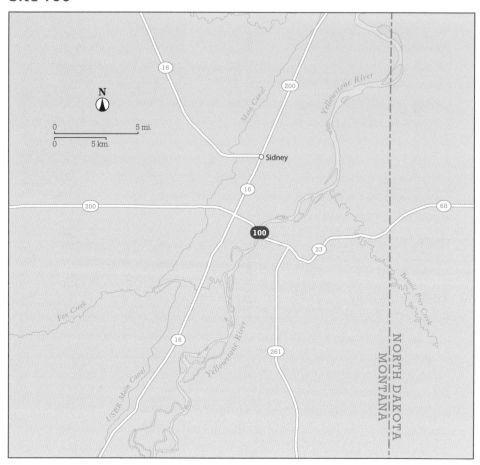

100. Sidney Agate

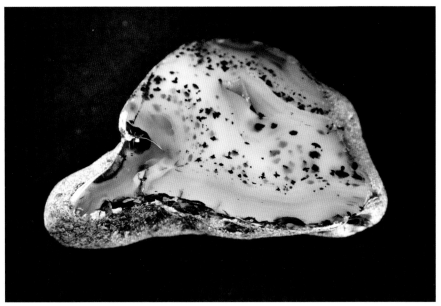

Cut and polished Montana agate at the Agate Stop

See map on page 216.

Land type: River, gravel bars

GPS: N47° 40.42920' / W104° 09.42420'

Best season: Spring through fall

Land manager: Montana Fish, Wildlife, and Parks

Material: Agate, petrified wood, jasper, chalcedony

Tools: None

Vehicle: Any

Accommodations: Camping, RV parking, and motels in Sidney

Special attractions: The Agate Stop, located on MT 16 in nearby Savage, is by far the most spectacular display and jewelry store for Montana agates in the state.

Finding the site: From MT 200 about 2.5 miles south of Sidney, turn left onto MT 23. Drive 1.5 miles, cross the Yellowstone River, and turn left into the fishing access. Park and explore the gravel bars.

Rockhounding

The best agate collecting in the Sidney fishing area seems to be upstream, where gravel bars are more exposed. This area has an abundance of material if you are willing to spend the time walking along the gravel bars. As you move farther upstream, the pieces become larger and more abundant. Most are translucent and many contain beautiful moss agate patterns.

APPENDIX A: GLOSSARY

agate: A variety of chalcedony closely related to opal that is often translucent and found in a variety of colors and patterns.

amethyst: A purple or violet variety of quartz often used as a gem.

ammonite: An extinct group of mollusks that may have been similar to the present-day chambered nautilus.

anticline: An upturned fold in the rocks of the earth's crust.

aquamarine: A blue-green gem variety of beryl.

azurite: A blue copper carbonate often associated with malachite.

Baculites: An extinct cephalopod that's similar to ammonites but is straight rather than curled.

barite: Barium sulfate occurring in blue, green, brown, and red colors.

batholith: A large mass of plutonic rock many tens of square miles in area.

belemnite: An extinct cephalopod possessing a cigar-shaped protective guard.

beryl: Beryllium aluminum sulphate that is generally colorless in its pure form; varieties include blues, pinks, greens, and yellows. Beryl constitutes such precious gems as emerald and aquamarine.

brachiopod: A marine animal with two unequal shells showing bilateral symmetry.

breccia: Angular rock fragments cemented into solid rock.

bryozoans: Small colonial animals that build calcareous structures in which they live.

cabochon: A gemstone that has been fashioned into a dome and polished.

calcite: A common crystalline form of calcium carbonate that is usually white or gray.

cephalopods: A class of mollusks that includes the squid, the octopus, and the chambered nautilus.

chalcedony: A microcrystalline variety of quartz.

chert: An extremely fine-grained siliceous rock exhibiting many colors.

columnal: A portion of the column, or "stem," of crinoids.

concretion: A nodular lumpy rock, generally sedimentary, that forms about a nucleus (often a fossil).

country rock: The common rock surrounding another deposit of material.

crinoid: A "flowerlike" echinoderm with a multi-armed calyx, or head, and a long column, or "stem," attaching it to the seafloor.

dendrite: A treelike pattern produced when minerals (usually oxides of manganese) crystallize in minute fractures in the rocks.

diastrophic: Natural processes that deform the earth's crust (tension and compression that produce faults, folds, etc.).

dike: An igneous intrusion that cuts across preexisting rock layers.

echinoderm: A phylum of marine invertebrates with spiny bodies, including starfishes, sea urchins, and sea cucumbers.

echinoids: A class of echinoderms that includes the sand dollar and sea urchin.

epidote: A green monoclinic mineral with crystals that are often used as gemstones.

fault: A break or fracture in the earth's crust along which movement takes place.

feldspar: The most widespread mineral group occurring in shades of white, gray, and pink in all kinds of rock.

fluorite: A clear-to-translucent mineral commonly occurring in shades of blue or purple as a common mineral in veins, which can be found in cubic crystals.

fossil: The remains of plants or animals preserved in rocks.

fossiliferous: Containing fossils.

garnet: Any mineral of the garnet group, commonly used as an abrasive. Garnet also occurs as a pink-red semiprecious gem commonly cut into stones.

gastropod: A type of mollusk with an asymmetrical unchambered shell.

gem: A general term for a variety of stones that can be cut for ornamental purposes.

genus: One of the divisions in the classification of living things (or fossils).

geode: A hollow nodule or concretion that may contain crystals.

geology: The science that studies the earth, its composition, the processes that affect the rocks of which it is composed, and its history.

geomorphology: The geologic study that deals with the shape of the earth's surface and the development of landforms.

gradation: A natural process such as weathering, erosion, or deposition that helps to shape the surface of the earth.

gypsum: A widely distributed mineral consisting of hydrous calcium sulfate that commonly forms in thick beds and occurs in shades of transparent (selenite), red, brown, and gray.

hydrothermal: Processes involving the action of hot-water solutions.

igneous: Rock that hardened from a molten state.

intrusion: Igneous rock that was intruded into preexisting rocks.

jasper: A variety of chert associated with iron ore that typically occurs in the color of red.

laccolith: An igneous intrusion that has squeezed between older rock layers and has domed up the overlying strata.

lapidary: The art or artist who fashions gemstones from rough rock.

limonite: A general term for a group of brownish iron hydroxide, commonly a secondary mineral due to oxidation of iron-bearing minerals.

malachite: A green copper ore that occurs in crystal and massive forms.

metamorphic: Rocks that have undergone physical and chemical change due to extreme changes in temperature and pressure.

mica: A group of sheet silicate minerals, major members of which include muscovite and biotite.

micromount: Small mineral specimens that have been permanently mounted and are best observed with the aid of a microscope.

mollusk: Any of the marine invertebrates in the phylum Mollusca, generally characterized by soft unsegmented bodies and hard calcareous shells, which includes cephalopods, pelecypods, gastropods, etc.

opal: A silicon oxide closely related to chalcedony. It occurs in a dull common form and the translucent precious form, which may contain flashes of "fire" and is used as a gemstone.

paleontology: The branch of science that deals with the study of fossils.

pegmatite: An exceptionally course-grained igneous rock with interlocking crystals usually occurring at the edges of batholiths; also found as lenses and veins.

pelecypods: A class of bivalve mollusks that includes oysters and clams.

placer: Generally a sand or gravel deposit containing minerals of economic value.

plutonic: Igneous rocks that hardened deep within the crust of the earth.

porphyry: An igneous rock with crystals of different sizes—usually large crystals surrounded by very small crystals.

pseudomorph: A crystal with the geometric form of a mineral that has been replaced chemically by another mineral.

pyrite: Metal-looking sulfides or disulfides. The brassy form is often known as "fool's gold."

quartz: Common rock-forming mineral composed of silicon dioxide. Crystals may be glassy or opaque (milky quartz) and exist in a variety of colors, including white, rose, smoky gray, and purple.

sedimentary: Rocks that form from the accumulation, compaction, and cementation of sediment.

septarian concretion: A concretion possessing internal fractures that are filled or partially filled with minerals.

siliceous: Containing silica.

sill: An igneous intrusion that parallels preexisting rock layers.

spirifer: A type of brachiopod possessing a butterfly-shape shell.

stratigraphy: The scientific study of the layers of strata in the earth's crust.

syncline: Down-turned folds in the rocks of the earth's crust.

thumbnail specimen: A small mineral specimen about 1 inch by 1 inch in size.

trilobite: Extinct marine arthropod that is beetle-like in appearance.

volcanic: Igneous rocks that hardened on the earth's surface.

vug: A small cavity in a rock.

zeolite: A general term for a group of hydrous aluminosilicates that can occur as well-formed crystals in cavities of basalt.

APPENDIX B: FURTHER READING

Alt, David, and Donald W. Hyndman. *Roadside Geology of Montana*. Missoula, MT: Mountain Press Publishing Co., 1986.

Bates, Robert L., and Julia A. Jackson. *Dictionary of Geologic Terms*. New York: Random House Inc., 1983.

Harmon, Tom. The River Runs North: *A Story of Montana Moss Agate*. Sidney, MT: Elk River Printing, 2000.

Korbel, Petr, and Milan Novak. *The Complete Encyclopedia of Minerals*. Edison, NJ: Chartwell Books Inc., 1999.

Montana Bureau of Mines and Geology. Various bulletins (www.mbmg. mtech.edu).

Prinz, Martin, George Harlow, and Joseph Peters. *Simon and Schuster's Guide to Rocks and Minerals*. New York: Simon and Schuster Inc., 1978.

Thompson, Ida. *National Audubon Society Field Guide to North American Fossils*. New York: Alfred A. Knopf Inc., 2001.

APPENDIX C: FINDING MAPS, LAND MANAGERS, AND INFORMATION

USDA Forest Service
Beaverhead-Deerlodge National
    Forest
Forest Supervisor's Office
Thomas Reilly, Forest Supervisor
420 Barrett St.
Dillon, MT 59725-3572
(406) 683-3900 or (406) 683-3913

Bitterroot National Forest
Forest Supervisor's Office
1801 North 1st St.
Hamilton, MT 59840-3114
(406) 363-7100 or (406) 363-7100

Custer National Forest
Supervisor's Office
1310 Main St.
Billings, MT 59105
(406) 657-6200

Flathead National Forest
Supervisor's Office
1935 3rd Avenue East
Kalispell, MT 59901
(406) 758-5200

Gallatin National Forest
Supervisor's Office
PO Box 130
Bozeman, MT 59771
(406) 587-6701 or (406) 522-2520

Helena National Forest
Supervisor's Office
2880 Skyway Dr.
Helena, MT 59602
(406) 449-5201

Kootenai National Forest
Supervisor's Office
1101 Highway 2 West
Libby, MT 59923
(406) 293-6211

Lewis and Clark National Forest
Supervisor's Office
1101 15th St. North
Great Falls, MT 59405
(406) 791-7700

Lolo National Forest
Supervisor's Office
Fort Missoula Building 24
Missoula, MT 59804
(406) 329-3750

Bureau of Land Management
Billings Field Office
5001 Southgate Dr.
Billings, MT 59107
(406) 896-5013

Butte Field Office
106 North Parkmont
Butte, MT 59701
(406) 533-7600

Dillon Field Office
1005 Selway Dr.
Dillon, MT 59725
(406) 683-2337

Glasgow Field Office
Highway 2 West
RR #1 - 4775
Glasgow, MT 59230
(406) 228-3750

Lewistown Field Office
Airport Road
PO Box 1160
Lewistown, MT 59457
(406) 538-7461

Malta Field Office
HC 65, Box 5000
Malta, MT 59538
(406) 654-1240

Miles City Field Office
111 Garryowen Rd.
Miles City, MT 59301
(406) 233-2800

Missoula Field Office
3255 Fort Missoula Rd.
Missoula, MT 59801
(406) 329-3914

Montana Bureau of Mines and
    Geology
Montana State University-Billings
1300 North 27th St.
Billings, MT 59101
(406) 657-2938

Montana Tech
1300 West Park St.
Butte, MT 59701
(406) 496-4167

US Geological Survey
Western Distribution Branch
Box 25286, Denver Federal Center
Denver, CO 80225
(303) 202-4700
(sells Montana USGS maps and
    publications)

APPENDIX D: ROCK SHOPS IN MONTANA

A&L Shoppers Pawn & Rock Shop
2101 Harrison Ave. #2
Butte, MT 59701
(406) 782-9000

The Agate Stop—Home of the
  Montana Agate Museum
124 4th Ave. North
Savage, MT 59262
(406) 776-2373

Crystal Hound
160 8th Ave. West North
Kalispell, MT 59901
(406) 756-2357 or (406) 212-2602

Crystal Imports
500 South Russell St.
Missoula, MT 59801
(406) 549-8907

Crystal Limit
2301 South Grant St.
Missoula, MT 59801
(406) 549-1729

Earth's Treasures
25 North Wilson Ave. #B
Bozeman, MT 59715
(406) 586-3451

Gold Miser
30525 US 2 South
Libby, MT 59923
(406) 293-8679
(prospecting supplies only)

Healing Wonders
442 Batavia Ln.
Kalispell, MT 59901
(406) 756-8705

Junction Rock Shop
7309 US 2 East
Columbia Falls, MT 59912
(406) 756-8705

Kehoes Agate Shop
1020 Holt Dr.
Big Fork, MT 59911
(406) 837-4467
(They sell rocks too.)

Montana Within Rock Shop
1761 Columbia Falls Stage
Columbia Falls, MT 59912
(406) 755-4788

The Prospector Shop
6312 US 12 West
Helena, MT 59601
(406) 442-1872

Rocks and Fossils
5170 US 89 South
Livingston, MT 59047
(406) 222-6725
(Also inquire about their other location, Dancing Bear Gallery of Gems, Minerals, Fossils, Jewelry and Beads Shop in Bozeman.)

INDEX

ABOUT THE AUTHOR

Partially due to her patriotic name, **Montana Hodges** has traveled throughout the western United States as a geologist and freelance writer. Her works
include a series of camping, hiking, and rockhounding books for FalconGuides.
She received her master's in science journalism at the University of Montana
and currently lives in Missoula, where she is working on her PhD in paleontology and anxiously awaiting the day royalties exceed expenses.